F. W Maurice

The System of Field Manoeuvres

Best Adapted for Enabling our Troops to Meet a Continental Army

F. W Maurice

The System of Field Manoeuvres
Best Adapted for Enabling our Troops to Meet a Continental Army

ISBN/EAN: 9783337405304

Printed in Europe, USA, Canada, Australia, Japan

Cover: Foto ©berggeist007 / pixelio.de

More available books at **www.hansebooks.com**

THE SYSTEM

OF

FIELD MANŒUVRES

BEST ADAPTED FOR ENABLING OUR TROOPS
TO MEET A CONTINENTAL ARMY

BY

LIEUT. F. MAURICE

ROYAL ARTILLERY

INSTRUCTOR OF TACTICS AND ORGANISATION, ROYAL
MILITARY COLLEGE, SANDHURST

WILLIAM BLACKWOOD AND SONS
EDINBURGH AND LONDON
MDCCCLXXII

NOTE TO READER.

This Essay was written in consequence of the following announcement, which appeared about a year ago in 'The Times:'—

"PRIZE FOR A MILITARY ESSAY.

"The Duke of Wellington, desiring to promote professional knowledge and the expression of original ideas among Officers of the Army, proposes, with the concurrence of His Royal Highness the Field-Marshal Commanding-in-Chief, to give £100 as a prize for the best Military Essay, on the following conditions:—

"'1. Subject of the Essay—"The System of Field Manœuvres best adapted for enabling our Troops to meet a Continental Army."

"'This subject will be treated under the following heads:—

"'(a) Mode of forming the columns of march when a collision with the enemy may be expected.

"'(b) Mode of covering an army on the march, or in

position, in order to conceal its movements, and to obtain information of those of the enemy.

"'(c) Mode of forming, combining, and employing the different arms for attacking an enemy in position.

"'(d) Mode of combining and employing the different arms for receiving the attack of an enemy.

"'Tactics have lately undergone, and may be expected to undergo, important modifications. The subject, therefore, opens a wide field for the display both of acquired knowledge and of original views. All theories or suggestions should be supported by argument, and, as far as possible, by precedents of recent warfare. Where general principles are laid down, the modifications which circumstances, such as different topographical conditions, may cause, should be explained.

"'3. No Essay is to exceed in extent 100 printed pages of the Queen's Regulations.

"'4. The competitors to be Officers on full pay or half-pay of her Majesty's Army, without restriction as to rank.

"'5. The Essays to be forwarded before the 1st of March next to Col. E. B. Hamley, C.B., Commandant of the Staff College, who has, at the Duke of Wellington's request, undertaken to decide among the competitors. Each Essay will be distinguished by a number and a motto, inscribed by its writer, who will also communicate the number and motto, together with his name, to the Duke of Wellington, at Apsley-

House, Piccadilly, to whom alone they will be known until the Prize is awarded.

"'6. Arrangements will be made for the publication (if it be deemed desirable) of the Prize Essay for the benefit of the writer, and his name will also be published.'"

The award on the then anonymous essays was announced in a letter from Colonel Hamley to the Duke of Wellington, which appeared in the 'Times' in May. As will be seen by the above conditions, the essays were sent in before March 1st. As, therefore, it was impossible that this paper should be made public till some months after it had been written, it seemed well both carefully to revise it, and to add references to a few works which had become available for use in the mean time. These works are noted in the list of books. A few references have been also added to older books, but as these were of no very serious importance, it has not seemed necessary to particularise them. Some delay has been inevitable in thus carrying out a revision of what had been laid aside for several weeks.

LIST OF BOOKS QUOTED.

This list is only intended for convenience of reference. As several of the works have been necessarily quoted many times, it would have been very inconvenient not to have had some shorter means of quoting them than that of giving the title-page. But the list is neither intended to imply that the writer has mastered the contents of every one of the works quoted, nor, as is in fact not the case, that they represent all those that he has consulted.*

TITLE-PAGE.	MODE OF REFERENCE.
Was enthält das neue Reglement? Gedrängte Zusammenstellung der abändernden Vorschriften welche im Neuabdruck des Exercir-Reglements für die Infanterie, d. d. 3 August 1870 enthalten sind. Von Rogalla-von Biberstein, Premier-Lieutenant im Zweiten Hannoverschen Infanterie-Regiment Nr. 77. Berlin, 1871.	"Was enthält" or "Was enthält das neue Reglement." *Note.**—The value of this book scarcely appears upon its title-page. The 'Neue Reglement,' though nominally dating from before the war, has in effect been only recently issued. It thus to a great extent indicates what changes the experience of the war has induced the Prussian

* It must be observed that both the above stand as they were originally. Colonel Newdegate's translation of the changes in Prussian drill has appeared since the essay was written; but it was scarcely possible to make the references to his translation correspond to those in the above little pamphlet, which was more convenient than the Prussian drill-book for my purpose, because it gave the essence of the changes in a short space.

	authorities to make. The book itself is simply a collection of these. It will be seen that the old form of skirmishing divisions had been abandoned before the war began.
Streffleur's Oesterreichische Militärische Zeitschrift. Wien, 1871. October. A review of the recent Prussian changes in drill.	Streffleur.
Do. November. A statement of the losses on both sides.	Do. November.
Do. December. Further statement of losses on each side in each action.	Do. December.
Das heutige Gefecht. Berlin, R. v. B. 1871.	Das heutige Gefecht.
Taktische Folgerungen aus dem Kriege, 1870-1871. Von A. v. Boguslawski. Berlin, 1872.	Boguslawski.
Tactical Deductions from the War of 1870-1871. By A. v. Boguslawski. Trans. by Colonel Lumley Graham, late 18th Royal Irish.	Boguslawski, Translation.*
Militarische Gedanken und Betrachtungen über den deutsch-französischen krieg den Jahre 1870-1871. Vom Verfasser des Krieges um Metz.	German General.
Opérations Militaires autour de Metz. Par un Officier Général Prussien. 1871.	Fighting round Metz.
Die Operationen der deutschen Heere von der Schlacht bei Sedan bis zum Ende des Krieges nach den Operations-Akten des Groszen Hauptquartiers dargestellt von Wilhelm Blume, Königl Preusz Major im Groszen Generalstabe.	Blume.
Do. do. Translated by E. M. Jones, Major 20th Foot, Professor of Military History, Sandhurst. With Map and	Blume, Translation.*

* These translations have appeared since the essay was written. It seemed more convenient for English readers to add to the pages of the original referred to when the essay was composed, the corresponding pages of those translations which have since appeared. I have also added references to the other new works, and to a few others in the above list to which my attention has been drawn during the revision necessary before publication.

List of Books Quoted.

Appendix. London: Henry S. King & Co., 65 Cornhill.	
The System of Attack of the Prussian Infantry in the Campaign of 1870-1871. By Lieut. Field-Marshal William, Duke of Würtemberg. Translated from the German by C. W. Robinson, Captain Rifle Brigade, Garrison Instructor at Aldershot. 1871.	Duke of Würtemberg.
Upon the Art of Operating under the Enemy's Fire with as little Loss as possible. By Major Tellenbach. Translated from the German by C. W. Robinson, Captain Rifle Brigade, Garrison Instructor, Aldershot. 1871.	Major Tellenbach.
Die Operationen der Süd-Armee im Januar und Februar 1871. Von Hermann Graf Wartensleben, Oberst im Generalstab. Berlin, 1872.	Wartensleben.
Operations of the South Army in January and February 1871. Compiled from the official War Documents of the Headquarters of the South Army by Count Herrmann Von Wartensleben, Colonel in the General Staff. London: Henry S. King & Co., Cornhill.	Wartensleben, Translation.*
Observations on the Influence that Arms of Precision have on Modern Tactics. By Field-Marshal Baron Von Moltke, in the 'Militair Wochen-Blatt' of 8th July 1865. Translated from the German by Lieut. H. R. G. Craufurd, R.A. 1871.	Von Moltke.
Der Deutsch-Französische Krieg, 1870, nach dem inneren Zusammenhange dargestellt, von A. Borbstaedt, Oberst z. D., Redakteur des Militair-Wochenblattes. 3 Lieferungen. Berlin, 1871.	Borbstaedt.
Do. do. Translated by Major F. Dwyer, Austrian Cavalry.	Borbstaedt, Translation.*
The War for the Rhine Frontier, 1870. Its Political and Military History. By W. Rüstow. Translated by Lieutenant J. L. Needham, R. M. A. 3 vols. 1871.	Rüstow.

* See note p. x.

Campagne de l'Armée du Nord en 1870-1871. Par le Général de Division L. Faidherbe. Nouvelle édition.	Faidherbe.
Elemente der Taktik aller Waffen, für Officier-Asperanten und jüngere Officiere von K. G. von Berneck. Berlin, 1870. Sechste Auflage.	Berneck.
About Tactics. By Captain Laymann. From the German edition of 1869. By E. M. Jones, Captain 20th Foot. 1871.	Laymann.
The Prussian Campaign of 1866, a Tactical Retrospect. Translated from the German by Colonel H. A. Ouvry, late 9th Lancers. 2d edition.	Tactical Retrospect.
On the Prussian Infantry, 1869. Translated from the German by Colonel Henry Aimé Ouvry, C.B., late 9th Lancers. 2d edition. 1870.	Prussian Infantry.
Précis of a Retrospect on the Tactical Retrospect and Reply to the Pamphlet on the Prussian Infantry of 1866. By Colonel Von Schellendorf. Translated by Colonel H. A. Ouvry, C.B., late 9th Lancers. 1871. 1st edition.	Colonel Schellendorf.
The Elementary Tactics of the Prussian Infantry. Translated by Captain Baring, Royal Artillery. 1871.	Elementary Tactics.
Regulations for the Training of Troops for Service in the Field, and for the Conduct of Peace Manœuvres. Translated from the German by Captain E. Baring, R.A. 1871.	Regulations for Training.
Details of Outpost and Patrol Duty from the Instructions for Prussian Infantry. By the late General Von Waldersee. 79th edition. 1871. Translated by Major-General Sir C. Staveley. 1871.	Sir C. Staveley.
Abrégé de l'Art de la Guerre. L. R. Rossel. Deuxième edition. 1871.	Rossel.
The Operations of War Explained and Illustrated. By Edward Bruce Hamley, Colonel in the Army, and Lieut.-Colonel Royal Artillery, &c. &c. 2d edition. 1869.	Operations of War.

List of Books Quoted.

Foreign Armies and Home Reserves. By Captain C. B. Brackenbury, R.A. 1871. — Foreign Armies.

The Military Systems of France and Prussia in 1870. By Captain C. B. Brackenbury, R.A. 1871. — France and Prussia.

Paper Read before Royal United Service Institution on the Position and Lines of Defence of the 5th Corps before Versailles during the Winter of 1870-71. By Major-General Beauchamp Walker, C.B. — General Walker (1).

Infantry Field Exercise. 1870. — Infantry Field Exercise.

The Soldier's Pocket-Book for Field Service. By Colonel Sir G. Wolseley. 2d edition. — Sir G. Wolseley.

Paper in Colburn's United Service Magazine, Jan. 1872. Standing Camps. By a Cavalry Officer. — Standing Camps.

Do., Dec. 1871. Our Cavalry System. By a Cavalry Officer. — Our Cavalry System.

Paper in Proceedings of the R.A. Institution. The Minor Tactics of Field-Artillery. By Lieut. H. V. L. Hime, R.A. (Royal Artillery Prize Essay of 1871.) — Lieut. Hime.

A Few Notes on the Handling of Horse-Artillery and Cavalry. By Captain Kitchen, R.H.A. — Captain Kitchen.

A Lecture given at the Royal United Service Institution. Practical Artillery. By Captain T. B. Strange, R.A. — Practical Artillery.

La Guerre de 1870. Par L. Vandevelde. 2d edition. — L. Vandevelde.

The Winter Campaign of Le Mans. No. 64, U.S.I. Papers. Captain C. B. Brackenbury, R.A. — Campaign of Le Mans.

Staff College Essays. By Lieut. Baring, Royal Artillery. — Captain Baring.

Reinländer Vorträge über die Taktik Gehalten an der K. K. Kriegschule. 8vo. Wien, 1870. — Reinländer.

Volunteer Discipline. By T. D. Acland, Lieut.-Colonel 1st Administrative Battalion Devonshire Rifle Volunteers. — Sir Thomas Acland.

Taktik nach der für die Königlich Preussischen Kriegschulen vorgeschriebenen "Genetischen Skizze" ausgearbeitet. H. Perizonius. 2 vols. Berlin, 1870.	Perizonius.
The Recent Campaigns in Virginia. Col. Chesney.	Recent Campaigns in Virginia.
Drill and Discipline. Captain Flood Page, Adjutant London Scottish. Henry S. King & Co.	Captain Flood Page.
On Outpost Duty. By Major-General Beauchamp Walker, C.B. A Lecture delivered at the United Service Institution.	General Walker (2).
Le Spectateur Militaire. 23d vol. Guerre de 1870. Par V. D., Officier d'Etat-Major.	V. D.
Les Causes de nos Désastres. Par un Officier d'Etat-Major de l'Armée du Rhin.	Les Causes de nos Désastres.
The Essentials of Good Skirmishing. 2d edition. By Colonel Gawler, K.H., late of the 52d Light Infantry. 1852. (1st edition, of which this is a Reprint, with Additions and Notes, published in August 1837.)	Colonel Gawler.
Avant-postes de Cavalerie Légère. Par le Général de Brack.	De Brack.
Aperçus sur quelques détails de la Guerre. Par M. le Maréchal Bugeaud.	Bugeaud.
Conférences Militaires Belges. Etude sur l'Emploi des Corps de Cavalerie au Service de Sureté des Armées. Par A. Fischer, Major au 2d Chasseurs à Cheval.	Conférences Belges.*
Conférence d'Etat-Major. La Cavalerie et son Armement depuis la Guerre de 1870. Par A. Lahure, Capitaine d'Etat-Major.	Conférence d'Etat-Major.*
Modern Cavalry: its Organisation, Armament, and Employment in War. By Lieut.-Col. George T. Denison, jun., &c.	Modern Cavalry.

* These papers reached England after the essay had been written.

List of Books Quoted.

Foy. Histoire de la Guerre de la Péninsule sous Napoléon. 4 vols.	Foy.
Studien über Truppen Führung, von I. v. Verdy du Vernois. 2 vols.	Truppen Führung.
On the Employment of Field-Artillery in Combination with the other Arms. By Kraft, Prince of Hohenlohe Ingelfingen, Major-General of the Brigade of Artillery of the Prussian Guard. Translated by Captain Clarke, R.A. 1872. (German edition dated Berlin, 1869.)	Prince Hohenlohe.
Souvenirs Militaires de 1804 à 1814. Par M. le Duc de Fezensac, Général de Division. Paris, 1863.	Fezensac.
Eine Militarische Denkschrift. Von P. F. C.	Prince Frederick Charles.
A Military Memorial. By Prince Frederick Charles. Translated by Charles Chambers, M.A. 2d edition.	Prince Frederick Charles. Translation.
Rapports Militaires. Ecrits de Berlin, 1866-1870. Par le Colonel Baron Stoffel, Ancien Attaché Militaire en Prusse. Troisième edition. Paris, 1871.	Baron Stoffel.
Paper read before the U.S. Institution. By Major Jones, 20th Regiment. "The Changes in Prussian Drill."	Major Jones.
Les Maréchaux de France. Etude de leur Conduite de la Guerre en 1870. Par H. Brackenbury, Captain de l'Artillerie Anglaise, &c.	Les Maréchaux de France.

The last book has not been published. It has been placed in my hands since the essay was written. Captain Henry Brackenbury, Royal Artillery, Military History Professor at Woolwich, was the principal representative in France of the Society for Aid to the Sick and Wounded. In that capacity he had opportunities for comparing evidence and sifting statements such as seldom fall to the lot of any student of war. French being as natural to him for purposes of composition as English, it seemed to him more courteous to the French army, with whom he had been so closely associated, to publish in that language, and in Paris, the upshot of his conclusions on the war. The above work, embodying these, had accordingly been announced, and would have some time since appeared; but, owing to some delays in the publication, it was not actually before the public when the French Government decided to try, at his own request, Marshal Bazaine by court-martial. It will not surprise English readers, though it appears very greatly to have astonished certain French newspapers, that under these circumstances Captain Brackenbury, at very great personal loss and inconvenience, thought it right to withdraw a book of which the title tells its own tale. As, however, personal matters in no way concerned my essay, and his evidence was more valuable as to certain facts than any other that could be obtained, he has kindly allowed me to quote freely from the book all that was relevant to my purpose, provided I did not put forward anything which should militate against the purpose which had induced him to suppress it. I have, of course, not put references in this case, but have quoted at full.

INTRODUCTORY.

MANŒUVRES have been defined to be "the quick orderly change of highly-trained and flexible masses from one kind of formation to another, or their transference from point to point of a battle-field for purposes which become suddenly feasible in the changing course of the action."* I have been led to the conclusion that the very basis on which at present our scheme for accomplishing this "manœuvring" is founded must be changed if we would meet the changed conditions of war. The objects to be attained are precisely those named in the definition: the method of securing them is greatly modified. Those who have been engaged in the recent fighting, and who have recorded their experiences, are very unanimous on the subject. I must therefore ask for a little patience, if, before proceeding to the detailed consideration of our future manœuvres under the several assigned heads, I am drawn into an inquiry, the relevancy

* Operations of War, p. 312.

of which will be, perhaps, not fully apparent till the details are discussed. As it is one which would not be relevant to the discussion of a system of manœuvres based on a drill dependent on prescribed words of command, it may seem at first to be disconnected from the issue raised by the proposed subject. I have urged throughout the following pages that the less we imagine we can dispense with any of the lessons of the past, the sounder our conclusions will be. But as I have maintained, nevertheless, the necessity for some very radical changes, it has been scarcely possible to approach the question from precisely the same side which would have been natural formerly.

GENERAL INQUIRY INTO THE NATURE OF OUR FUTURE MODE OF CONDUCTING BATTLES, &c.

Is it necessary carefully to reconsider the principles on which our mode of fighting is at present based? Must we go through them all, and, testing them by recent experience and probable analogy, decide how far each of them continues to be applicable to modern conditions of war? Or will it be sufficient to modify a few details, and to continue to apply all our old maxims as of yore?

To reconsider and re-examine is not necessarily to reject. But the necessity for the larger inquiry appears to be forced upon us by a glance at even the barest outline of recent events.

It was essentially the tactical facts of the late campaign in France which were calculated to startle every soldier interested in his profession. The war of 1870, no doubt, in all its circumstances, singularly falsified the prediction of the statesman who had warned us at its commencement "not to expect dramatic catastrophes." But with regard to the broader features of the campaign, what chiefly added to the startling and dramatic effect was our ignorance of the antecedent conditions.

That an army whose organisation had decayed should have been utterly beaten by one led by more able chiefs, numerically very superior, perfectly prepared for war, does not now cause us much astonishment.

Nor were soldiers in general particularly surprised to have it proved for the hundredth time that armed mobs are not armies—that when once the trained forces of a nation have been crushed, the civil populations, however great their enthusiasm, and however vast their numbers, are not likely, unassisted, to offer a successful resistance to invasion.

When, however, we turn to the accounts of the various actions of the war, the case assumes an altogether different aspect.

Under all previous conditions of tactical action, the army which surrounded another, without a most unusual numerical superiority, had exposed itself to the risk of disaster, if not of destruction. No very recondite reason lay at the bottom of this fact. The surrounded army could always attack its enemy at some point with superior numbers. Yet, almost to the last man, the last horse, and the last gun, the French army passed into captivity because every portion of it had been in succession literally surrounded.

Moreover, no one of the capturing armies had over the forces which surrendered to it such an advantage in numbers as would, according to all previous calculation, have justified its extension over the enormous expanse of ground which it occupied at the moment of victory.*

Nor were the features peculiar to each of the three main incidents of a kind to render the facts less startling. Among the French troops at Metz were those *élite* corps for the sake of which all other regiments had been denuded of their best men. On the whole, the conditions of topography and the ample if not adequately armed defensive works appeared to favour the besieged. They were commanded by a general who, in whatever qualities he may have been deficient, had certainly not previously been supposed to be below the average of commanders either in talent or in recklessness. It is clearly not a sufficient answer to say that critics dispute the necessity of Marshal Bazaine's comparatively passive attitude during the siege, and that Frenchmen assert that he might have cut his way through on the 31st August. According to all our

---

* Blume, p. 224 (trans., p. 246), for instance, gives the space occupied by the Prussian army during the sortie of the 19th January from Paris at 20,000 men for 8000 paces—5 men to 2 paces; about 5000 men to the mile in all. Compare also Duke of Wurtemberg, p. 36, 37.

previous notions, it ought not merely to have been possible but easy for him to have done so. Certainly at the present moment, taking into account all that has been said by Bazaine, by his French assailants, and by the Prussians, there is at least very serious doubt whether he could have broken through at all after he was once fairly "surrounded."*

Neither does the case of Sedan present a less serious puzzle. Assume that the French surrender was due to mere panic and disorder; how does that account for the conduct of the Prussians? Before they could have been at all certain of the condition of the French—instead of hemming them in upon the neutral territory, as, according to all precedent, they ought to have done, they preferred to send troops round between the enemy and Belgium. They chose, that is, to effect the capture by what a short time ago would have appeared to all military students the ridiculous method of literally surrounding them, rather than by that practical military process which consists in cutting off all chances of escape.

---

* I refer less to his own "Rapport" than to the discussions in the 'Revue des deux Mondes,' continued throughout '71 repeatedly, and to the 'Fighting Round Metz,' which is mainly devoted to a careful consideration of the question.

Since the above was written the new pamphlet has appeared. I suppose that no one can read the reports addressed to the Marshal as to the events of the 31st without coming to the conclusion that if it was possible to have overcome the German resistance so completely as to have broken the line before it was reinforced, at least that operation encountered difficulties such as it would not have met with under any previous conditions of war.

It may be true that the actual position occupied on 1st September was the result of the gradual development of the drama. But if any one who knows something of the previous history of attempts to "surround" armies will take in hand one of the many maps showing the situation of the two forces on that day, I think he will agree with me in this conclusion. A general so able as Count Moltke would under no previous conditions of war have allowed the course of events to lead him into exposing his army to the risks of such a situation. When we know who the leader was, the mere map, with the positions of the troops marked on it, is wellnigh sufficient proof that a vast change has taken place in the application at least of our present principles.

Nor despite all the disorder of Paris can that case be considered less startling. M. Thiers had in 1842 persuaded the French people to expend what is said to have been then nearly two-thirds of the value of all real property in the city upon its fortifications. He was supported in his demand by the most experienced soldiers of the world. They certainly were not then under the mere belief that he would thereby render the capture of Paris difficult, but under the firm conviction that he would render it under any circumstances utterly impossible. It was not the first time that Paris had been exposed to the danger of a siege by an army of 250,000 men. Nor were there many more trained soldiers in Paris in 1814 than in 1870. There was nothing, therefore, very novel in these conditions. For such circumstances M. Thiers and his advisers ought to have been prepared by experience. But what they had not believed was that an army such as this could extend itself over so many miles of ground, and yet present at each point any serious power of resistance. Nor can it be doubted that, if the wisest and most experienced officers of any army in the world had been asked a few years ago if such an extended line could possibly be maintained, the answer would have been in the negative.

To these facts it must be added, that of all European armies, that which succumbed under these conditions was the one which by tradition and training was the most likely to take full advantage of the special fault of over-extension on the part of its enemies; and that, on the whole, the armament of the victors was inferior to that of the vanquished. The value of the potential numerical superiority at any one point, which was possessed at all events by the army of Metz, ought to have been enhanced by the latter cause.

In the presence of such events as these, it is not wonderful that we should be exposed to the risk of falling into one of two opposite errors. Now, as always, there are, of course, those who will maintain that whatever is, is right; who will take "the form for the essence," who will insist that we should "rigidly adhere to maxims and traditions long after they have ceased to be applicable;" and the "pedantic spirit which blindly confides in them, and condemns all innovations, is by no means yet extinct among soldiers."* However serious the risk on this side may be, it would scarcely be useful to meet it here by argument. Those who adopt this line repudiate all argument, and they must be allowed to elect for themselves the method in which they are to be treated. On the other hand, however, some of our very ablest soldiers are inclined, not perhaps unnaturally, to adopt language which is fraught with at least as great a danger. We are in presence, they say, of altogether new conditions of warfare. All the past must be forgotten. We must start from first principles; we must deduce these from what we know of the breech-loader, and—I am sorry I cannot put the idea more respectfully—"from the depths of our internal consciousness."

I have no wish to underrate the importance of the present crisis. I am not urging that any principle that we have adopted from our experience of the past should be held to be sacred simply on that account. The experience of the past on which it is founded must be carefully examined. Whenever we find that the conditions have so changed that the evidence is no longer good, it must be ruthlessly rejected. All modern circumstances must be carefully taken into account. No

* Operations of War, p. 324.

maxim or tradition ought to be allowed to stand, simply because it was applicable formerly.

But no one who has listened frequently to the conversation of many of our ablest and most thoughtful officers, can fail to have been struck sometimes by a tone which certainly does not promise the most careful examination of the whole case, or the best result finally. If an illustration from any war previous to that of 1866 be quoted, its value is instantly disputed, not because the circumstances which have changed have in the least degree affected its relevancy, but on the broad assumption that "you can't draw any deductions from what occurred before the introduction of the breech-loader." The question is, On which side does the burden of proof lie? To me it seems to be altogether on the side of those who introduce each particular novelty. If I am asked to state what in fact does continue to be valuable in the experience of the past when so much is changed, it might be sufficient to throw on those who ask the question the notoriously difficult task of proving a negative. But I feel more inclined here to take out of many a single illustration, which, at a time when skirmishing has acquired its present importance, will scarcely be undervalued. There is an army which can boast that it once contained a division which acquired an experience of eight years' continuous skirmishing, without the completeness with which the experience was applied and adapted being once impaired by the division ceasing to exist as a military unit. That experience has been in no small degree recorded. Charges of plagiarism are of course absurd where kindred experience leads to identical statement. No one would be disposed to accuse either Prince Frederick Charles, Boguslawski, or Captain May, of enacting the part of Mr Puff. But I shall have occasion hereafter to note that in the case of each of these great soldiers, with reference to many of their most admirable suggestions, "The same idea occurred to two men," and an English light-infantry officer " thought of it first."* If that were all, the matter would be one of personal or literary interest; but in the pages of Colonel Gawler's book will be found not a few practical hints, palpably unaffected by subsequent changes, which have at least been not so fully brought out in any German work on the late war which has yet reached England. Of course many parts of Colonel Gawler's pamphlet are no longer applicable. The question is, Shall we, because of these, reject the experience of those that are? And, which is more important, shall we apply to all the rest of our accumulated store of tactical experience the principle of careful selection or of reckless rejection?

At all events, those who assume that they can start with an absolutely clear field labour under two very serious disadvantages. In the first place, the ground

* See title-page in list of books, "Colonel Gawler."

is, as a matter of fact, very well filled already; and if they ignore the existence of the old plants never so much, the roots will encumber the ground and interfere with the growth of their young saplings. If it be only for the sake of eradicating the old ideas, they must accept the fact of their existence. If it is ignored, these ideas will be constantly in practice the source of fresh deductions, not modified at all by more modern circumstances. But secondly—and I confess to thinking this much the more serious matter of the two—it is by no means evident that the experience of 1870 and of 1866, supposing the latter campaign to be accepted as at all in the same category, is sufficiently ample to admit of correct and complete conclusions being drawn from it.* That it is sufficient to show the necessity for an immediate and careful examination of our present position I have already maintained. That very many absolute deductions may be made, and that the general drift of all the changes that will be ultimately necessary may be surely gathered, I firmly believe; but I fear we must accept the fact as inevitable, that for some

---

* As the German General has pointed out, p. 245, only two battles on a really grand scale, Gravelotte and Konig-gratz, have been fought out since breech-loaders were introduced; in one of them they were employed only on one side. It is a curious fact, too, that on almost the only point on which Captain May and Colonel Schellendorf were agreed—the disappearance of key positions on battle-fields—the German General differs from both of them. "Who shall decide when" such "critics disagree?"—Conf. Prussian Infantry, p. 81, and Colonel Schellendorf, p. 33, with German General, p. 245.

time after each new chapter of the history of war has been commenced, tactics are unsettled as to their details.

It will be some time yet before all the facts which are likely to present themselves under the new phase of war have so occurred and been so recorded that we shall be able to eliminate the purely tactical causes from a variety of other operative circumstances which now tend to confuse our judgment. We may be able to arrive at principles not discoverable at first, but it will be long yet before the new mode of warfare approaches perfection.

It seems, therefore, more important at present to avoid a too hasty assumption that anything is proved for which we have not very ample evidence, than to render the study as complete in detail as it can be made. That progress is surely safest which "broadens down from precedent to precedent." The precedents of the past must be so considered that their relation to present facts may be discovered. The action of the present on the future must be taken into full account. That is not a reason for assuming the existence of evidence we do not possess.

It becomes, therefore, important for us to inquire first, what evidence we have on which we can safely rely as to the changes which new conditions have introduced into the art of manœuvring armies. It will then be necessary to consider further, how far the evidence is applicable to the present or possible condition of our army, and how far, therefore, it is essential that the condition of our army should be modified to meet the new facts. Before examining under their several heads the different branches of the art, it appears to be advisable also to treat under the general title those conditions which are common to the whole system of manœuvres, because the relation of several of them to one another is so intimate that it will be scarcely possible to examine them separately till their connection with one another has been taken into account. It will be hard, for instance, to decide on the right method of conducting a march till certain facts have been studied with reference to the modern circumstances of fighting. Yet in point of time the marching precedes the fighting, and largely influences its character.

First, then, as to the evidence. We certainly do not possess that best of all kinds of evidence which alone can be considered decisive of the questions at issue, "the practice of good modern generals in forming troops for defence or attack." * For, as is plain on the face of the case, the formations which were adopted by either army during the war are very defective evidence as to the nature of those which the best generals of that army would adopt after the experience gained in the course of the struggle. The French system of manœuvres failed as palpably as any other part of their scheme of war. The German formations for the

* Operations of War, p. 329.

attack and defence of positions were developed by the stress of events.

It may be almost said that as evidence of the formations which their leaders would now select in the presence of a powerful army, those actually adopted by the Germans are the less trustworthy of the two. For as the experience of the Germans increased, so their need of employing it with care diminished. It is impossible to assert that even their ablest chiefs had fully thought out the whole problem before the war began. The attack on St Privat la Montagne, and the order which followed it, that the experiment was never to be repeated, are conclusive on that subject, at all events.* Moreover, the order implied a change in the very principles which, up to the last moment, had been instilled into the minds of all young officers. The pages of Berneck's sixth edition, published at the beginning of the year of the war, which treat of the bayonet attack (page 34, &c.), and in fact the whole series from page 30 to page 40, might have served almost as an account of the grounds on which such an attack as that on St Privat ought to be made. The absolute order against its repetition implied, therefore, that the German generals recognised that even the principles on which they had based their

---

* Compare Duke of Wurtemberg, p. 18, with Borbstaed, p. 358 to 362; Translation, p. 460-464. I shall hereafter have occasion to quote from 'Les Maréchaux de France' certain facts which intensify the importance of the tremendous loss sustained by the Guards very greatly.

training required serious modification.* The perfection with which an army had been prepared for war, which could thus in the immediate presence of the enemy adapt itself to events, is all the more striking; but it is impossible not to infer also that the formations actually adopted were due rather to the instinctive aptitude of the army than to the deliberate determinations of the chiefs. Almost immediately after the battle of Gravelotte had been fought, all necessity for further developing the system of manœuvring in presence of highly-trained troops passed away for all the Prussian armies except that which encompassed Metz. The motley and demoralised host which surrendered at Sedan was scarcely in reality more formidable than the raw levies whom the Germans had subsequently to meet.

It is scarcely necessary to urge that it is impossible to judge of the manner in which German leaders will in the future conduct operations from their manner of fighting against Gambetta's "armed men." If any

---

* On the other hand, it must be remembered that the new Prussian "regulations for drill," admirably adapted as they are in all essentials to present conditions, though issued only in '71, were ready in August '70, a few days before Gravelotte. See, however, 'Streffleur,' October. The necessary forms had not been adopted. The spirit was ready for them. Berneck says expressly in this sixth edition, page 39, that any more skirmishers than can be helped are "rather mischievous than otherwise;" but in the 'heutige Gefecht' (like Berneck, intended for young officers), published since the war, skirmishing is as completely everything as it is for the German General. The elaborate systems of Perizonius and Reinländer, issued just before the war, show no signs of the change.

doubt had remained on the subject, it would be removed by the recently published *quasi* official narrative of Major Blume. There is a tone too pitiful to be ever contemptuous about the manner in which the narrator refers throughout to the comparative ease of the latter portion of the work of the German armies.

On the whole, therefore, the following deductions appear to be legitimate. At the commencement of the campaign the German leaders had not been prepared to adopt those modifications in tactics which were proved to be necessary almost at once by their first experience of fighting against the breech-loader and the mitrailleuse. Such modifications as were subsequently introduced, so far as they proceeded from the orders of the generals commanding, were therefore necessarily only tentative. They had not been carefully tried and studied in the school of mimic war. Moreover, they were never perfected even under the more rapid tutorship of battle itself. The need for perfection passed away. The time came when the only thing that was dangerous was not to dare enough. We cannot, therefore, judge absolutely of the future practice of the German generals from either period of the late war. For in the first portion they had not learnt the necessities of the new conditions of things ; in the latter, other circumstances had rendered even these new conditions of comparatively secondary importance.

This, then, the best of all evidence, being unattainable, on what are we to base our conclusions? We have little choice. If it were possible to obtain a complete statement from various witnesses of the several phases of the actual fighting of the late campaign, and of the success or failure which attended various experiments, it might be possible from these to get sufficiently clear light as to the proper conduct of future battles. Unfortunately for any such complete collection of facts as would be useful in this respect, we shall still have long to wait. It has been happily said, "If you want to obtain any such ground to work upon, you must put off your attempt till the regimental histories of the war are published. Read those between the lines, and you will find what you require." As, however, it is impossible to delay so long before deciding upon the broad principles at least of our future action, we are almost compelled to obtain our statements as to facts from the little better than casual illustrations introduced by those who would defend a particular theory.* The regular narratives of the battles contain so much that is irrelevant to these tactical discussions, so little that gives one a clear picture of the manner in which the fighting really went on, that very few practical deductions can be made from them. Fortunately, the nation which is at present most deeply engaged in the discussion, is of all the one which loves to pay the most plodding devotion to fact, and to re-

---

* I refer more particularly to "the German General," Boguslawski, the Duke of Wurtemberg, 'Das heutige Gefecht,' Major Tellenbach.

quire the most thorough investigation before accepting any statement. Hence it happens that into almost all the argumentative treatises which have appeared since the war, a very solid substratum of evidence as to fact has been introduced, by which one is able largely to test the accuracy of the conclusions of the authors. On the whole, the points on which most of these writers agree as matters of general experience, may be assumed to be tolerably established, more especially since they have been largely confirmed by foreign military critics. Only it is essential to remember that the ground on which we are working is of an altogether different kind from the distinct positive study of such clearly-recorded battles as those of Austerlitz, Waterloo, or Solferino.

Such evidence as was afforded by the latter with reference to the phase of tactics which they illustrated, we do not with reference to the present possess. Nor is it only that the facts of the present campaign will not for a long time be so sifted as to supply us with a clear narrative of what occurred. When they are so sifted, the evidence will not be of at all the same kind as in the other case. At Austerlitz a perfected system displayed its power against an effete one, while yet the excellence of the opposing troops made it necessary that all possible force should be developed to crush them. At Waterloo the attack and defence were conducted respectively by the masters who had severally perfected each in that age of war. At

Solferino the blunders that were committed were open, palpable, and tended distinctly to make more clear the rules of tactical action which were violated. I have already alleged the grounds on which it seems to me that the circumstances of the late war present nothing of similar definiteness for tactical study. In obtaining such facts as there are, we have no doubt the advantage that many of our own officers were present during the later period of the war, and have contributed very valuable hints as to the events that took place.* But this period is that which is least valuable for tactical study ; and on the whole, for the mass of our evidence we are obliged to rely on the statements of those German authors who adduce the facts expressly in order to establish their own theories. The French at present seem not to have realised the extent of the change that has passed over the nature of fighting. Even Rossell speaks of the late campaigns essentially as if they had been of the same character as those of the past in their tactics as in their strategy, or had proved only the advantage of a war of American shelter-trenches.† The majority of

---

* I refer rather to private conversations than to public utterances, but more especially among the latter to the Lectures on Le Mans, and letters in the 'Times,' of Captain C. B. Brackenbury, R.A. The paper in the R.A. Institution's Proceedings, by Colonel Smithe, R.A., is almost the only personal evidence by an English officer as to the early part of the war which we possess. I may now add Captain Henry Brackenbury's invaluable book, should it ever be made public.

† On which subject see 'Das heutige Gefecht,' p. 7.

French military writers, mistaking utterly the form for the spirit, quote Jomini in a manner that might well bring the old strategist from his grave if they had not heaped over him so heavily the dead relics of his maxims.* I am very far from meaning that we have not much to learn from these writers; but we certainly have not had from that side, as yet, any comprehensive view of present tactics.

I urge thus strongly the necessity for considering carefully what the sources are from which our knowledge of the tactical facts is derived for this reason. We no longer have before us, as we had in the case of the great battles of the past, the whole of the facts of the case, or at least as nearly the whole as we can obtain by comparing many different narratives, simply as narratives. We are hearing a statement which appears complete from the writer's own point of view. We are hearing it, moreover, not as would be the case with a simple narrative of facts—in such a manner as to show us clearly the extent of the writer's personal opportunities for judging—but as if the statement embraced the whole field of all observers. Now, for foreigners in particular, there

---

* I refer especially, however, to the elaborate 'Guerre de 1870, par V. D.' Both it, Vandervelde, 'Les causes de nos désastres,' and not a few others, are very suggestive as to certain matters of detail. I am not speaking in the text of the generals, but of the military writers. Chanzy and Faidherbe, under the most difficult circumstances, have surely given evidence of very far-seeing views. At present, however, Chanzy withholds all comment.

seems to me to be danger in too readily assuming that, however able and fair the writer may be, he has put the whole of the facts before them. It is before all things necessary, under such circumstances, to take account of the writer's whole position, both personal and national, in deciding how far the exact form in which his evidence is given conveys its real meaning to ourselves. I am not at all disputing here the wisdom of the propositions maintained in these various works; but on all grounds it is necessary that we should be first clear what the actually proved facts are before we accept deductions from them.

My point will become clearer if I begin by considering the first German study of modern conditions of war, which excited great interest in England, and which, however incomplete it may necessarily have been, certainly showed a foresight, scarcely, to say the least of it, surpassed even by the leaders who conquered France.

When the 'Tactical Retrospect of the War of '66' first appeared in England, it was erroneously supposed to have been either written or inspired by Prince Frederick Charles. That it represented much that most soldiers who cared for and studied their profession had been long thinking, was recognised at once. Men hailed with applause the General who had so thoroughly penetrated the conditions of the time, and had been so able to divest himself of the special prejudices of his own position as to see the necessity for developing the tactical independence of action of *captains of companies*. Is it saying too much to assume that a very considerable proportion of the value of the evidence disappeared when the work turned out to have been written not by Prince Frederick Charles but by Captain May? I do not, of course, mean that truth is less truth wherever it comes from. Nor do I argue that such a question as this ought to be decided by authority rather than by reasoning based on experience; but the question is as to the nature of the evidence before us. Now there cannot be much doubt which is the stronger evidence as to the real facts of the case. A captain asserts, from what he has himself seen on the field of battle, that the sphere of a captain's authority ought to be very greatly extended. Is it quite conclusive? Had a general been struck by the same fact, would not the probability that he was right have been indefinitely stronger? Nothing impresses one more in reading all the theories that are propounded by different men, than the extreme difficulty with which any one emancipates himself from the prejudices of the special position in which he happens to be placed. Nevertheless it would be idle to deny that Captain May had hit upon a truth in his assertions under this head. It has been brought out more clearly by all the circumstances of the late war even than by the expositions of the writer. It was acknowledged to be true in principle by the very German authorities who most opposed the exact form in which the writer expressed it, and the excess to which he carried it in one special direction.* It was accepted almost by acclamation by the whole military public of Europe. But the truth was a larger one than Captain May had quite realised. It had touched him on one side, and there he had seen it. But the stress of battle is too severe to allow men to be well aware of what concerns others rather than themselves. Where the same truth applied to other ranks, and distant points of the field, May had not realised it.

It is worth while to follow this out very fully, for much hinges on it, and on the character which must be impressed on our whole system of manœuvres by whatever solution is arrived at, of the questions which Captain May was the first to bring prominently forward. Captain May asserted that the whole scheme of manœuvres, in presence of an enemy, adopted by an army acting on the offensive, must be based on the greater independence of action left to the captains. It was a bold assertion; for it suggested a revolution, not so much in the mere rank in the military hierarchy on which authority would pivot, as in the whole idea of manœuvring at all. Hitherto our system of manœuvres has been based on a system of drill. That drill has consisted in a series of elaborately regulated changes of formation, chiefly from line to column and from column to line.

* See especially the preface to Colonel Schellendorf's paper.

As a kind of appendage to these we have recently added, for all our infantry, what used to be the *spécialité* of a particular branch, a system of somewhat looser drill. The principle of the whole drill has, however, been carefully to train all subordinate commanders to know the exact words to give on receiving the command of the drilling officer, and the men to know exactly in what way to move on receiving the executive command. By this means it has been possible for the general to prescribe the exact nature of the movement to be performed by every section of his army on the parade-ground, in order that he may be able to retain the same power when in presence of the enemy. Captain May's assertions, if they be accepted as true, would sweep away the whole of this idea. For, however necessary it may be to retain rigid drill as a means of discipline—as the first step, that is, in training each man to take his proper place in the whole organisation—certainly, as a means of teaching men the principles which are afterwards to be applied before the enemy, it will no longer be applicable. Yet, however numerous may have been the points on which those who have followed Captain May have differed from him — in whatever things he was clearly proved to be wrong in the answer that was made to him by authority—certainly few will now be prepared to dispute that he was right in this broad principle with reference to the future character of manœuvres. The one point on which it is scarcely possible to find now any difference of opinion among those who have most carefully studied recent facts is, that the distinct formal movements by which an army is made to assume throughout a similar or a corresponding formation, are no longer possible in presence of modern weapons. It is absolutely essential, in order to diminish the disastrous effect of the present arms, that each small section of an army should be moved in such manner as the local circumstances impose.* Each body, acting, so far as detail is concerned, independently, must be able to modify its formation so as to take the utmost possible advantage of favourable conditions of ground, and to act as the conduct of the enemy and the relative effectiveness of his fire from time to time suggest.

On the other hand, while Captain May erred rather in restricting the application of this principle too much than in extending it too widely, he affords himself an admirable illustration of the necessity for that superior care and supervision the importance of which he almost wholly ignored.† It was possible for so

---

* Major Tellenbach throughout; Duke of Wurtemberg, p. 28; German General, p. 234, 235; Boguslawski, p. 54, 66, 80; Trans., p. 58, 71, 85, and throughout repeatedly; 'Das heutige Gefecht,' p. 4.

† In 'The Prussian Infantry' Captain May seems at times to be arguing on behalf of the very authority on which he pours contempt. But there can be little doubt, as Colonel Schellendorf assumed, that the only ranks the necessity for which he strongly felt were those of the general, the captain, and the private. The reserve in the hands of the general is his one idea of any other action than that of the captain's.

keen and intelligent an observer to imagine that because the range of weapons and their deadliness of effect have been together immensely increased, that therefore the necessity for men who shall pay attention to larger considerations than those which present themselves to the view of a captain has passed away as soon as battle has commenced. How very limited must have been even his perception of the dangers to which his own company and the whole army would frequently, had he obtained what he wished, have been exposed! Surely when sudden development of overpowering fire from some unexpected quarter is infinitely more probable than it was formerly, the importance of the work of those who are to provide for such contingencies is infinitely increased, and not diminished. When a captain's duty is to pay the closest possible attention to the conditions of ground where his company is, and to the local action of the enemy, he will be less, not more, able than he was before to judge of the larger features of the contest. But if, accepting Captain May's evidence as good as to what he saw, more especially in so far as it has been confirmed by writers on the present campaign, we modify our deductions from it by making allowance for his actual position, it is not at all difficult to see that the conclusion which we ought to draw from

All other ranks, under his arrangements, seem so utterly in the way, that one wonders why they should be put there at all, unless it is from a courtesy which puts them on a pedestal to be laughed at.

it is not different from that deducible from later facts. It is when he comes to the arrangements of the battlefield on the grand scale that he appears to err, from taking as the basis of all his calculations less a complete investigation into the duties and the importance of each rank than his own partial experience. It is one proof amongst a thousand of the errors which arise from a confusion between the duties of the witness and those of the jury. Captain May claims that we should accept his conclusions as inevitably correct, because he has himself evidence of value to bring forward. In practice it is when he steps from the witness-box to that of the jury, and still claims to speak with the authority of a witness, that he commits the blunders that have been exposed by his no less experienced critic; so at least Colonel Schellendorf appears to have felt.

It will be necessary, therefore, to consider further some of this more recent evidence before drawing out absolutely the conclusion to which, as I believe, a proper examination of May's brilliant pictures of events would lead us, just as a study of the later facts also does.

All writers who have seen anything of recent fighting appear to be agreed that an attack in column is a thing no longer possible. On this one point we do, moreover, possess the best of all possible evidence—the practice of the generals who have most successfully applied modern arms to warfare. For the order

Future Mode of Conducting Battles. 29

issued after the attack on St Privat la Montagne was not founded on any possible anticipation of the feeble resistance which the Prussian troops were subsequently to meet with, but on direct experience of the loss sustained in that attack.* Some mystery, no doubt, hangs over the exact circumstances of the disaster. "Some one," pretty obviously, "had blundered;" but the details, whatever they may have been, cannot affect the significance of the prompt action taken at headquarters.

An attack in rigid line, except for short distances, never was possible against properly-posted enemies.† Attacks never can now, except under the rarest circumstances, be restricted to short distances. The change which has come over fighting in this respect is in fact due to three main causes: first, to the new necessity for bringing out the efficiency of the weapon in offence; secondly, to the impossibility of facing modern fire in any close formation whatever;‡

---

* The abandonment by the Prussians of the special form best known in England of even their company columns (see 'Was enthält,' &c., p. 5) points in the same direction, showing, as the Duke of Wurtemberg also does, p. 29, that even small columns, as such, cannot advance now against a position. German General and Boguslawski so assume always.

† Operations of War, p. 378; Von Moltke, p. 15.

‡ The following letter from a staff officer who was present at the incident described, was recently read by Captain H. Brackenbury at the United Service Institution :—

"Le 16 Août, vers une heure à la bataille de Rezonville, le Commandant du 6 Corps révenant de la droite qui occupait St Marcel vers la gauche qui s'appuyait à la route de Verdun, à quelques centaines de

thirdly, to this, that the development of the power of weapons would demand for line-attacks a perfection in drill unattainable by the best troops at a moment when it has quite ceased to be possible to restrict the training of troops to drill. At all events, other questions apart, I think that conclusion must be forced upon all who study Count Moltke's criticisms on the advance on the Alma. That in some form or other, therefore, if an army is to retain *the*

mètres en avant de Rezonville, vit de l'autre côté de la route, un mouvement de retraite s'operer dans le 2ᵉ Corps et un mouvement d'attaque se prononcer par l'infanterie Prussienne sur le hameau de Flavigny. Il prescrivit aussitôt au 94ᵉ de ligne de se porter sur Flavigny, puis autant pour soutenir le moral de ses troupes en presence de la retraite des troupes voisines que pour aider à arrêter le mouvement de l'infanterie Prussienne sur Flavigny il porta en avant deux bataillons déployés du 93ᵉ que précédaient des tirailleurs à 3 ou 400 metres. Masqués jusqu'à ce moment par un pli de terrain, ces deux bataillons à peine découverts, furent en but à un feu d'artillerie tel, qu'ils purent à peine franchir quelques centaines de mètres et qu'ils fléchirent sous le feu qui les écrasait, et avait en quelques minutes, mis plusieurs centaines d'hommes hors de combat. Ce feu provenait de la grande batterie établi par l'ennemi en arrière et au sud de Vionville, c'est à dire à plus de 3000 m. de nous, et qui, suspendant son feu sur Flavigny l'avait concentré sur les bataillons, au moment où leur mouvement en avant avait été distingué. Cette batterie était forte de 24 pièces suivant les uns, de 36 suivant les autres; une partie des pièces étaient couvertes par un épaulement. En ce moment l'infanterie Prussienne était environ à 2000 m. de nous entre Vionville et Flavigny et son feu ne nous causait aucun mal. Les bataillons du 93ᵉ qui prirent part à ce mouvement comptaient un grand nombre de jeunes soldats, puisque l'avant-veille de la bataille ce regiment avait reçu à Metz un contingent de 900 hommes qui n'avaient jamais touché un chassepôt. Neanmoins ces bataillons ne reculirent pas sous le feu, mais fléchirent sur eux-mêmes, et je doute qui des troupes plus vieilles aient pu prolonger leur mouvement en avant beaucoup plus longtemps que nos jeunes soldats, qu'entrainait lui-même notre commandant de corps d'armée."

power of attack at all, it must nowadays attack in skirmishing order, with a proper system of supports and reserves, seems scarcely to admit of a doubt.

Now the very fact of skirmishing implies an increase in the space of ground occupied by a given number of men. Nor is this the only cause which tends to make the space covered by specific numbers much greater than it was formerly. No doubt it is easy to lay too much stress on the evidence afforded by the extreme extension of the German formations during the later period of the war. It is impossible, however, to ignore the assertions of a recent writer, who combines singular modesty of statement with indisputable ability, and who has had exceptional opportunities of observation. The Duke of Wurtemberg maintains that the space taken up throughout the recent campaign by small bodies of Prussians was by no means due merely to their contempt for their adversaries, but that, on the contrary, the very severity of fire induced men instinctively to spread in order to pass off towards those points where fire was least intense. This spreading appears to have been both local—small bodies extending out towards the parts of a position which proved to be weakest—and also, on the larger scale, to have taken the form of a constant tendency to those flank-attacks which were so marked a feature of the larger tactical movements of the campaign.

It inevitably follows, that as the space covered by

each hundred men is greatly increased, it becomes impossible for one officer to bring the same number under his eye as formerly. At the same time, the necessity for local adaptation to ground renders it no longer possible to keep men in formations in which they can at all moments see and attend to the immediate command of a distant officer. Yet rapidity of action has become more than ever essential, since a few minutes lost in conveying an order may sometimes now imply the destruction of thousands. To this it must be added that the horrible noise of breech-loader fire prevents distant orders from being heard, and that the intensity of each man's personal absorption in the work immediately before him under the new conditions tends to the same result.*

Yet the whole army, and each section of it, will be exposed far more than formerly to dangers not dependent merely upon the local circumstances to which the attention of subordinate officers is necessarily restricted. That Jomini's well-known argument as to the uselessness of getting a comparatively small body on the flank of a large army will be much less invariably applicable than it was formerly, is at least suggested by more than one incident of the late war.† Nor is it difficult to see why this should be so. A small number of men who are able

---

* Boguslawski, p. 71; Trans., p. 77.

† In Truppen Führung, ii. 25, there are some excellent remarks on the subject, though deduced only from '66 experience.

to bring an effective fire to bear upon a large body hampered by the necessities of manœuvring, will have a much greater advantage over them than they formerly had. For, during every moment lost in falling back to a new formation, many more shots will be poured in by the outflanking body than would formerly have been the case. At the same time, the very effectiveness of the assailants' fire will render the operation of falling back more difficult, and therefore the re-establishment of order will take longer. So that there are three things to be taken into account— the increased time which it will take to re-form, the increased effect produced by the assailants' fire within a given time, and the increased distance to which it will be necessary to fall back in presence of the greatly-extended range of the weapon. It seems not unreasonable to infer that that will almost always happen which occurred so frequently during the late war—the whole defence will collapse as soon as even small bodies of assailants appear on the extreme flank. But in proportion as a flank attack thus assumes even more than its former deadliness of character, an attack which breaks the line will produce also greater effect than ever. For now, as always, to break a line will be to threaten two flanks of the enemy. Hence it is essential that provision should be made to resist any attempt either to penetrate the line or to outflank and roll up the army. For the skirmishing body being constantly liable to over-extension, while yet most of its efficiency depends on not locally limiting the degree of extension it assumes, the difficulty must be met in some other way. The need at each link throughout the army—the corps d'armée, the division, the brigade, the regiment, the battalion, the company—of some such provision, might, I believe, be illustrated by actual incidents of the late war.

Taking, then, the whole of these facts into consideration, the inference does not appear difficult to draw. It is this: For an army that would be able to meet *all* the circumstances of present warfare with the same freedom with which the Prussian has met them, the first great necessity is, not that the free action of the captains should be exceptionally developed, but that the free action of every rank, from the general to the private, should be fully developed,—not in order that each rank may interfere with and claim independence from the rank above it, but in order that each may more effectually co-operate with and carry out the work assigned by that immediately superior to it.* All training must tend to develop

---

* The difficulty of giving references in this case to the recent writers is, that one would have to quote nearly every page of each book—'Das heutige Gefecht,' German General, Boguslawski.

If the deduction was, however, not plain on the face of it, it would be sufficient to refer to pages 2 and 6 of the 'Regulations for Training,' and still more forcibly comparing these with the new drill regulations, it is most striking with how much increased intensity the authors assume that the character of the training is not a question of improving the effectiveness of an army any longer, but is the essential principle on

the qualities which are essential to such a manner of action. The habit of command must cease to be the habit of exact prescription, and become the habit of clear instruction.* Men must be constantly accustomed to act under orders which they will have to interpret according to circumstances; otherwise, when they find themselves under the necessity of deciding, they will think it essential to decide absolutely for themselves, instead of deciding how they can best carry out the views of those who command them. Unity or harmony of action will be more essential than ever, but it must be arrived at by a thorough appreciation of the spirit rather than by a strict adherence to the letter.

At the same time, it is clear that a definite means of exercising influence on the field of battle must be specially provided for each rank. For it is necessary to meet that difficulty which has been pointed out by Captain May and others—that from

which manœuvring is based, and on which the character of our system of manœuvres depends. So also in our own drill-book, "Manœuvres of troops in the field represent the application of the principles acquired at drill."—'Infantry Field Exercise,' 279. Obviously, therefore, the drill must be adapted to the principles which ought to be taught, or manœuvring becomes impossible. Compare also Operations of War, p. 416, and Truppen Führung, ii. 26.

* I must point out here that my consideration of the subject is solely limited to the *tactical aspect* of training.

I am not discussing those larger questions which have been so eloquently dealt with by Captain C. B. Brackenbury in 'France and Prussia in 1870;' nor even all with which Major Jones deals in the lecture on Prussian drill, some of which, if it were possible to go into them in such an essay as this, touch the subject very closely.

the moment that any portion of an army is thrown forward into actual fight, it ceases to be under the hands of any but its most immediate officers. There seems but one way in which to do this—to attach to every rank a special reserve, to be kept in hand till emergency requires its use; the effect of even a small body of fresh troops suddenly thrown into the fight being, according to all testimony, as striking as is the extent to which those actually engaged pass out of hand. I would, then, propose that there should be special company reserves, special reserves for the battalion and for the brigade, as well as for the division and the corps d'armée, and independent of any general reserves that may be retained for the whole army. In this way it will be possible, without serious risk, to permit much greater independence in detail to each subordinate, without trenching upon the authority and influence of the superior. It is clear enough that the character of the work which will be required of each rank will be infinitely higher, more difficult, more indicative of skill, than it ever has been before. Those above the rank of captain have no occasion to fear lest they should find their work taken away from them, as Captain May declared that it was. On the contrary, they will be relieved of details, but will have far more important matters to attend to.

Such seem to be the legitimate conclusions to be deduced from the statements and arguments of Prussian writers, and from the obvious character of the late fighting. Before proceeding to apply them, however, it is necessary to consider how far they can be applied to our own army. For here, also, it is important to remember how largely the manner in which various writers put things is influenced by their special position. It is at least as necessary for us to take care that, in this instance, we do not forget the national starting-point of the writers, as in the case of Captain May it was necessary to remember his personal position. Accepting as true the assumption that, for the future, freedom of manœuvre will depend on the manner in which authority is delegated from hand to hand, and on trained habit rather than on rigid conformity to rule, we are brought at once to this. The radical change which has taken place in tactics is, as it was at the time when the system of Frederick gave place to that of Napoleon, one before all things in organisation.* But the change is an infinitely more vital and complete one now than then. If a flexible chain was then substituted for a bar of iron,† it remained dead metal still, more pliable under the hands of the one man who wielded it; each link capable of a certain degree of independent *motion*, but essentially it was intended to obey only mechanically the impulse that was imparted to it. *We* have to provide a new substance. A living organism has to

---

* Operations of War, p. 324.
† Operations of War, p. 333; quotation from Baron Ambert.

take the place of a material instrument. It must work under the inspiration of the regulating head, rather than move with mechanical precision in the directing hand. If, therefore, our army does not possess such an organisation as will enable it to meet the new condition of things, or cannot have such an organisation adapted to it, it is useless to inquire what, in the abstract, is the best system of manœuvres which we could employ against a Continental army. It will be far better to take, not the best, but the second best, or whatever it may be, which we can effectually employ in our actual condition. To attempt the manœuvres which would be suitable to an army capable of freedom of action, while we are entirely incapable of it, would be as wise as for a dwarf to go to battle with the weapons of a giant. Yet in war there is no "best" but victory, no "second best" but defeat. It may be questioned, therefore, whether the adaptation of our army to present tactical necessities is not a matter on which its continued existence should depend. To what, then, has the immense facility of manœuvring which the Prussians have shown been due?—first, doubtless, to the perfection of the actual training for war which has been acquired severally by each man throughout the army.* A system of working at once

---

* This is put so forcibly by L. Vandevelde that I quote it here, though, as will be seen hereafter, I do not admit that it expresses the whole truth: "En Prusse chaque soldat est un petit tacticien et à plus forte raison tous les officiers le sont. Qu'on ne s'y trompe pas : on n'est pas tacticien parce qu'on sait se déployer en tirailleur sur une plaine

so free and harmonious would have been impossible if all had not been trained to appreciate the value of the same principles, and to understand the larger theory of the great art in the details of which they had to co-operate. At every point the training of the average Prussian officer shows itself to have been as high as it is probably possible that, for the ordinary run of an army, it ever should become, whether in the practice or in the theory of their profession. But was that all? I cannot think that any one who has considered the history of the camp of Boulogne,* and the effect which it had upon the succeeding wars of the Empire, and who studies the features of the present war, remembering always what is the nature of the German organisation in peace time, will doubt how important an element that permanent local organisation of the corps d'armée has contributed to the marvellous harmony of their tactical working. So much admira-

unie, ou former une compagnie à droite ou sur la droite en ligne ou en bataille. La tactique consiste à savoir tirer un bon parti d'un terrain à connaître quelle est la formation qu'il convient d'y adapter dans les diverses circonstances de combat, et sous ce rapport le soldat comme l'officier Prussien ont une instruction tactique que les officiers et les soldats des autres nations n'ont pas reçue."—P. 101, 102.

* Operations of War, p. 326. Fezensac (p. 31) has shown conclusively that, great as the gain in efficiency due to the camp was, that result was in no sense brought about by drill. Ney, the commander of the corps d'armée, held only two great manœuvres one year, none the next. The general of division had only three bad divisional drills. There was no brigade drill whatever. But Fezensac has no doubt as to what was *the* gain. "Le plus important de tous fut de s'accoutumer à vivre ensemble, d'apprendre à se connaître," p. 35. The whole passage is too long to quote, but well worth study.

tion has been justly expended upon one feature of that system with which the present essay is not concerned —namely, the extraordinary facility for rapid mobilisation which it offers—that it seems to me that the effect which the same cause has had upon the freedom of German tactics has never been sufficiently brought out. Yet, if an eleven of cricketers which has played habitually together meets another, the members of which are fairly equal to the other set in individual skill, but of which each member sees every other on that cricket-field for the first time, is it an element of trifling or rather of paramount importance in our calculations as to the probabilities of the result? If the management of a great railway company were suddenly handed over to officials, each of whom was individually skilful, but who had never worked together before, would any insurance company venture to guarantee the lives of passengers? Were formal drill and exact prescription adequate means for preparing an army for present warfare, and did they represent the principles on which it is advisable to manœuvre in presence of the enemy, then no doubt it would be wholly unnecessary that the several grades of officers who have to co-operate together should be personally and intimately acquainted. For a man at present receives with equal facility from any commander the fixed words which indicate to him those which he is himself to issue. But if the possibility of continuing our present system in presence of the enemy has passed away, and it is necessary that subordinates should intelligently co-operate with their superiors, then it is essential that the men who are to work together should not become acquainted for the first time at the moment when war breaks out. The higher the ranks the more essential is it that they should have been accustomed for a long time to trust one another. The German writers do not very pointedly draw attention to the effect of this feature on the campaigns they have passed through. Why should they? They write not for us, but for themselves. They have lived and been brought up under a system which makes a corps d'armée almost as much a family as a regiment is among ourselves. They probably, moreover, do not fully realise how great the advantage is to them, for they never knew what it was to be without it. But the more one studies the nature of the orders issued from one officer to another during the late war, the more one sees how the intimate personal knowledge that each had of the other enabled the exact amount of liberty that was required to be accorded, without preventing the fullest instructions from being given wherever they were really needed. Moreover, the habit of intrusting details to subordinates, and the habit of having details left by superiors to be worked out, is fostered in superiors and subordinates respectively by acting with men who are known to be trustworthy. Hence when the inevitable gaps of war do come, the injury is infinitely less than it would otherwise be; for the principle is established, and new men drop into their places very much more quickly where the nucleus is at each moment habituated to work together than where all are strangers.* I am obliged here to appeal to what is the common experience of all under kindred conditions. The facts are patent. The Germans do show this harmony of action. They do possess this special preparation for it. Since, as I have noticed, it is in the last degree improbable that they will ever point out to us themselves how all-important is the connection between the two, may we not fairly assume it on the *a priori* grounds here put forward? I fully admit that the essential thing is the thorough appreciation throughout the whole body of the right principles on which work ought to be carried on, and habitual training in the application of these to special ground and circumstances. But you will never obtain these by mere preaching. The question is, how to make them a reality? I confess I see no other way than the one I have indicated. For, before all things, it will be

* Faidherbe has collected a vast number of the orders issued to the Northern German army (p. 102-120). They all the better illustrate my point, because Göbcen was brought to the command from another corps. The habit established among an army is not shaken by necessary changes, especially where the men appointed are known in reality very well to the army. The contrast, however, to, say, Napoleon's army of 1815, where all had been tried in war, but never together, is too striking not to suggest its cause to every soldier. See also Manteuffel's Orders, especially Wartensleben, p. 39, 40, 41; Translation, p. 48, 49, 50.

necessary, among the many forms of free action which must be intrusted to commanders of each rank, to intrust them with entire freedom in restricting on special occasions, or on every occasion, the liberty of action which they accord to their subordinates.* There are times in war when everything turns upon the question whether the one exactly right thing is or is not done even by some small body of men. Now the commander who is perfectly conscious of seeing distinctly under such circumstances the exact detail which ought to be prescribed to subordinates, must be as little chained to the new rule of habitually intrusting details to them as to any mere rule whatever. Reproaches may justly of late have been cast against Napoleon† for his tendency to absorb into his own hands all initiative, till his "chain" almost became "iron rod" again in its stiffness, without recovering cohesion. But it is impossible not to admit that many a victory was gained by that superabundant energy which substituted, in the execution of details at essential moments, the genius of the great commander for the talent of

---

* The new Prussian drill regulations seem to contemplate this, where it is laid down that the latitude allowed to subordinates is never, *except under the most pressing circumstances*, to be withdrawn from them. —Was enthält, p. v.

† I refer, of course, to the whole train of recent assailants, but Rossel has perhaps put it, all things considered, as savagely and as neatly as anybody: "La décadence de l'art militaire en France date de Napoleon. Ce génie entier et jaloux ne voulant pas de rivaux, ne forma pas d'élèves."—Rossel, Preface, p. 1.

some fair subordinate. Men who are capable of seeing when such exceptional action is necessary, must be permitted freely to adopt it. But it will be fatal to allow that to become habitual which ought only to be considered a breach intended to honour the rule. How can both these objects be attained otherwise than by the habitual association of men in large masses where individual idiosyncrasies are recognised and corrected? while long habit of working together gives a force throughout each link to authority of the only kind which will not be shaken by the inevitable break-up of accustomed forms in presence of the enemy.

One of the ablest writers on the late campaigns, Boguslawski, has declared that disorder so inevitably, under present conditions, supervenes soon after fighting has commenced, that the only possible course is to accept the fact and "order disorder." In other words, he proposes what Captain May had previously urged,* that an officer should habitually take command of any men, no matter of what regiment, whom he finds around him dispersed by the circumstances of action. It is clear that

---

* Colonel Gawler having as usual anticipated them. "The true summit of perfection in skirmishing is the preservation of order in disorder, and of system in confusion." "In hot contests over large extents of intricate ground, men of different companies, regiments, brigades, and even divisions, mingle with each other. Soldiers should therefore be drilled, not indeed to fall into such irregularities on principle, but to be ready for them in practice." And much more of the like, *written in* 1837, too valuable to be taken out of its context.

enormous force will be lost, unless the disordered masses of various regiments, which, according to all testimony, now gather after a position has been taken, can in some way be led on to fresh victory. But the success of the proposed remedy depends on its being properly adapted to certain facts of human nature which it is all-important to take into account.

Under the Prussian system, which the writer of course assumes to exist, it is not at all difficult to understand that the men of two regiments which have been lying side by side one another, in Bonn or in Cologne, for instance, for years, would be easily trained to follow almost as readily the officers of the other regiment as of their own. Nor would it be difficult to extend the principle to the whole of the regiments whom they see habitually on the grand parade, and whose officers they have been for years bound to know and to salute. When once the habit has been established, on grand divisional field-days, of acting after attack under the command of any officer who happened to be up at the moment, it would scarcely require any great stretch of the same principle, even if, as the author says, corps d'armée become intermingled, and the officers of one corps d'armée have to take command of men of the other. But would it be possible, without any analogous training, to trust that the men of a regiment which had just arrived from Tipperary, and had never taken words of command from any but their own officers, would satisfactorily follow those of another which had just landed from Bombay? I cannot bring myself to think so. Even with all the conditions at present existing in the Prussian service which favour such a mode of fighting, Boguslawski considers it necessary not only thus to "order disorder," but to "practise disorder." I am quite ready to admit the immense advantage which we possess in the nature of the tie which binds together our officers as a class and our men as a class. I believe it to be at this moment at once the most organic relationship—that is, the one in which each class best understands what the nature of the relationship is—and the most cordial, hearty, and friendly existing between an upper and a lower class anywhere in Europe. Nor is it possible to deny that at Inkerman our men did fight nobly in numberless cases where they were necessarily commanded by officers of other corps. But the case rather serves to support the point I am maintaining than to weaken it; for the troops who fought at Inkerman had been for months in the same camp. Had this not been so, the circumstances of the case by no means presented those temptations to break away from authority which, according to all testimony, the conditions of modern fighting inevitably do. All men of common-sense, even if unaccustomed to discipline, when they have to fight like rats in a hole, instinctively place themselves under some one's orders. It is a very different thing to follow an unknown leader in a fresh movement, the nature of which is not realised. It is too firmly settled a conviction of my own mind for me to have any sense of national boastfulness in saying it, that if any troops in Europe can be trusted to do this thing without previous training, then ours can. But I doubt if any army in Europe would do it. At all events, that clearly was the conclusion as to our army of an old Peninsular officer not wont to depreciate it: "Soldiers who have not been drilled on this principle," says Colonel Gawler — and be it remembered he has already spoken of the intermingling of men of different *divisions*, the largest organic unit of our army in those days — " or who have not acquired it by experience, are, when extended under fire, continually liable to be transformed into unmanageable mobs."—Colonel Gawler, p. 15.

The causes which make this the fundamental question of all, in considering the nature of our future tactics, have been by no means yet all considered. The necessity that theoretical as well as practical knowledge should be universal throughout the army, scarcely needs to be proved. The advantage which a body of men possess, all of whom thoroughly understand and accept the same principles, is, as I have noticed, written on every page of the history of the late war. Yet, more than ever, the art of war is a constantly progressive one, based on the experience of the very latest as of all the past, reaching forward into the future as fresh inventions have to be studied and their possible applications considered. How, then, can an army, for all practical purposes isolated in battalions, be provided either with the instruction or with the implements and means of instruction, which are needful?

Moreover, at a time when it is rather the spirit of the present phase of fighting than knowledge of any special forms which can be instilled into the minds of men, there is another fact which is of vital consequence. Men are always infinitely less conservative of what requires to be changed; yet infinitely less ready to throw away all the experience that has been acquired from the past, when thought passes freely throughout large numbers, where methods different in detail are observed and compared. In small societies every private crotchet is apt to reign supreme, while new light is hardly admitted, for there is no neighbouring region whence it can enter.

Yet again, the less merely formal drill becomes a final and adequate preparation for the manœuvres of war, the more elaborate must be the training by *practice*, in order that each rank may be accustomed freely to adapt itself to the orders it receives. By no other means can mere looseness as opposed to elasticity be avoided. This, however, bears so closely on the subject that it requires fuller treatment. In the new modifications of Prussian drill, the principle laid down as the one thing vital, even for the company, is, that it shall be able to assume formations readily, for which it has not been previously prepared by training. The words are these: "The company must not only be so trained that the captain may have it well in hand, but that it may also be capable of executing, when ordered, movements previously *not much practised*"* (Was enthält, p. 4). Moreover, it is expressly laid down that the latitude allowed by regulation to subordinate officers is never, except under the most urgent circumstances, to be withdrawn from them. The effect of these regulations, in giving greater elasticity and adaptability to movement, is obvious at once. The one means by which the several grades are kept in the hands of the superior, is by pressing upon each the duty of passing as soon as possible again under the effective orders of their several commanding officers. But if we adopt this principle, or something analogous to it, how changed will be all the special points to which both captains and battalion commanders have to direct their attention! The very data on which a colonel was able to count securely before, will no longer be even elements in his calculation. The space occupied by battalions and by companies will no longer be fixed, but variable. The battalion commander will need all his at-

---

* "Nicht *besonders* eingeübt" (Was enthält, p. 4). Colonel Newdigate, however, translates an apparently similar passage, " which *has not* been (previously) practised"—rather an important difference. I have not seen the original, except in the pamphlet quoted above. There can, however, be no doubt that the passage, taken as a whole, conveys the stronger meaning which Colonel Newdigate gives to these words.

tention to see that he is not cut off from neighbouring forces, that his line does not snap where the enemy may threaten some too overstrained company, that he is ready to support at the right moment some all but successful attack. The same principle will of course extend itself upwards. That it can be met only by a fully-developed system of successive reserves, I have already urged. But how constant will the necessity become for practising large bodies together, if such a system as this is ever to reach perfection! The more we have to trust to the aptitude rather than to the memory of every individual soldier and officer, the more essential will it be not to allow immense discrepancies to occur. It will take far more frequent practice to insure aptitude than to cultivate memory. To prevent freedom of manœuvring from degenerating into incoherent independence and eccentricity, will be no easily accomplished task. Now the aptitude which must be both developed and regulated, consists chiefly in attaching their proper value to local circumstances, yet in not sacrificing to these what is necessary for perfect co-operation with others. How can this be adequately acquired except by men who are accustomed from time to time to work together in large numbers?

It by no means follows that the greater portion of drill should consist in such large manœuvres. Rigid formations will still be a most essential means of early training, and be also best adapted to most marches out of the immediate reach of the enemy. It is important that any details that can be suppressed should be done away with, in order that troops may be able to devote as much time as possible to perfectly mastering those which continue to be practical, and to acquiring field aptitude.* But always enough will remain to demand much time. No one who has watched the effect of much loose work upon ill-trained troops, will doubt that as a means of discipline parade drill will be more,† not less, essential than ever, little as it continues to be applicable to the purpose for which it was first designed.‡

The training of the individual soldier cannot any

* Operations of War, p. 416; Was enthält, as to recent changes in these respects, the suppression of various details connected with company columns of countermarching, &c., and in broad terms the bold assertion that the manœuvring facility required is not to be secured by memory.

† According to the latest reports from Berlin, the Prussian Guards are now largely practising even advances in line, and firing in line, which they never attempted before. The motive is obviously to give the troops the most difficult training possible, that they may be perfectly in hand.

‡ Boguslawski, however, maintains, surely with reason, that what he calls "pure tactics"—that is, tactics unadapted to ground—are, as a means of instruction, irrespective of discipline, no longer applicable at all; and therefore, that drill applied to the ground should be looked upon as the principal part of the whole, the other being only necessary for any infantry as a preparatory training.—Boguslawski, p. 151; Trans., p. 163, &c. Prince Frederick Charles had anticipated him to a considerable extent, p. 34, and Colonel Gawler had equally anticipated both, p. 33—a fact, though only one out of many to which I would venture to direct the attention of the worthy alderman and member of Parliament who recently, in a speech to his constituents, disputed the value of autumn manœuvres, on the ground that no soldier had urged their necessity till they were introduced by Mr Cardwell. I fear that he would find it a severe punishment to have to read the number of works in which they had in fact been asked for.

longer be allowed to consist even mainly in the mere acquisition of drill. It is essential that, since soldiers, despite every precaution, will be much less under the absolute control of either non-commissioned officers or officers than they formerly were, and will therefore have the power of constantly creating disorder, that they should learn how inestimable are the advantages of order. Every possible means must be taken to make them wish for victory, and to see that victory is a result that can only be acquired by the perfect co-operation of all ranks. The means by which all this can best be done, do not concern this essay; they have been admirably discussed in numbers of German works.* It is only necessary here to observe that, on the extent to which we can succeed in this matter will depend the kind of manœuvres we are able to adopt.

Moreover, now more than ever the success of the larger manœuvres will hinge on the extent to which the smaller units of an army are possessed of a training adapted to suit them. During the grand manœuvres themselves, it is rather, as it is happily said in the Prussian drill instructions, the senior officers than the subordinates who acquire training.† The placing

---

* Notably Laymann, Regulations for Training, Prince Frederick Charles, p. 31, 32, 35, 36; Trans. p. 45, &c. Those who think that the *first* thing of directly military importance in skirmishing is to develop the soldier's *confidence in his weapon*, will find their view fully confirmed by Colonel Gawler (p. 28 and p. 51), from Peninsula experience.

† Though, oddly enough, Prince Frederick Charles takes just the opposite view. Conf. Instructions for Training, p. 5, and Prince Frederick Charles, p. 28; Trans., p. 42.

of a company perfectly in the hands of its captain must be done chiefly by company training; a similar result must be secured for the battalion by its separate training.

On these three, no doubt—the individual training, the company, and the battalion training—the larger proportion by far of time must be expended. But unless the habit is acquired, by at least annual manœuvres, of looking upon all these as only preparatory to the larger work—and all of them are adapted to meet those necessities which are realised only when working with much larger bodies—how can aptitude, as distinguished from a knowledge of drill, be made to pervade all ranks? Thus some means of occasionally so manœuvring becomes essential, not as a specimen test for a selected portion of the army, but as part of the necessary training of every battalion, gun, and squadron in it. The less one particular locality is the scene of the manœuvres, and the more the instruction afforded by these is extended to each district in the country, the more valuable will the instruction be.

To sum up. Our manœuvring in the field can no longer be regulated by a system of prescribed words of command. Its precision, its harmony, and its success will depend instead upon a certain trained aptitude for working together acquired by the whole army, and by every individual in it. This aptitude cannot be developed unless in some way or other those men who in war time are to work together have been as a rule accustomed to work together in peace time. This also is more, not less, necessary, because it will be essential that men who have worked little or not at all together before, should in emergencies be able to work freely together. There the absolute tactical deduction stops. In the complete application of these principles, other considerations, with which this essay is in no way concerned, have to be taken into account. It is obvious that an extreme difficulty presents itself in the application in detail of the local corps system to England. The German armies of defence and of offence differ little in size from one another. With us the case is far otherwise, and this and various other matters determine the exact form in which the tactical result can best be secured. But the tactical necessity that the men who have to co-operate with one another in presence of the enemy shall have worked together beforehand, applies to each particular army that may be engaged in war. It is a matter altogether distinct from the question of the administrative convenience of local organisation, whether as to effective peace service or rapid mobilisation. To bring it back to the definition of manœuvres with which this paper starts, a general now, as formerly, requires that his troops shall effect "quick orderly changes" "from one kind of formation to another," and wishes to transfer them "from point to point of a battle-field for purposes which become suddenly feasible in the changing course of the action." He will be able to attain his wishes or not in proportion as his troops have become "flexible masses" by virtue of this previous habitual association and *this* kind of "high training," which consists as much in a prepared harmony of action in unforeseen circumstances as in a knowledge, spread throughout all ranks of the army, of the principles on which the mode of meeting such circumstances ought to be determined, and a practical readiness to apply them as events present themselves.

Certain modifications of minor organisation seem to be needed, partly in order to enable troops to adopt the formations required by the new arm, partly in order to develop as far as possible the habit of independent action among subordinate commanders, so that they may not fancy, when they are forced by stress of battle to decide for themselves, that the duty of conforming to the spirit of their instructions no longer binds them. A more elastic but a stronger bond must, by these means as well as by others, be substituted for that which we have—not because it is less essential than formerly that the whole army should be closely bound together, but because there is constant danger lest the bond should snap.

From this general point of view it appears very desirable to enlarge the size of our companies. For since the circumstances of fighting tend more and more to make cohesion depend on the instinctive readiness of men to follow out indications given them rather than on conformity to words of command, it becomes desirable that those who give the actual leadership and guiding in detail should be as far as possible the officers who are thrown into the most intimate personal contact with the men. Now, exaggerated, and to some extent flippant, as much that Captain May said on the subject of the respective relationship of captains and of colonels with their men undoubtedly was, there was this much of truth in it— in every army it is the captain who is immediately responsible for that minute detail which alone necessitates constant personal intercourse with the men. This fact, moreover, tends to increase in importance as the instruction of the men becomes of higher character, and therefore more and more requires the personal attention of officers rather than of non-commissioned officers.* The increasing extent of ground occupied by a given number of men tends also to make the company a more convenient body than the battalion for direct command. But it is obviously inconvenient that numbers so small as those of our present companies should be formed into distinct tactical bodies.† Their size has not been fixed with

---

* It seems to me to need no proof, that if our future fighting is to be of the kind I have maintained that it must be, then the instruction of the men must be superintended by company officers, not by adjutants or musketry-instructors. Analogous changes will be required in each of the other arms. It does not follow that specially-instructed officers will not be required to keep up the standard of knowledge among officers and non-commissioned officers.

† I use the expression " distinct tactical bodies," not "distinct tacti-

any object of the kind. Clearly, also, it will be almost impossible for the commanding officer of a battalion to look after a very large number of independently moving companies. The system of moving by wings has apparently not succeeded during the late campaigns. Hence it would seem well very largely to increase the number of men in a company, and to diminish the number of companies in a battalion.

Again, the great difficulty in the proper use of the new arm lies in the constant necessity for preventing waste of ammunition at unprofitable distances, and under unprofitable circumstances. Despite all the care they have expended in the individual training of their men, the Prussians have found it indispensable, in order to check the constant risk of waste, to do all they can to place the soldier as completely as possible in the hands of his officers as to the number of rounds he fires, the time of firing them, and the object aimed at.* It must be remembered that though the number

---

cal units," advisedly; because I think the inevitable deduction from the discussion on the subject of "units" between Colonel Schellendorf and Captain May is, that it is one of those "conclusions in which nothing is concluded." What we really require is, that each body should be so perfectly built into every other, that at each successive stage of the building a perfect unit shall be formed. A brick is not less complete because it is arranged in a course, or a course less regular because it is built into a wall; nor does a wall less perform the special part assigned to it because it forms one of the sides of a house. A quotation in 'Standing Camps,' apparently from a private conversation with Prussian officers on the present condition of our army, suggests the above illustration of patent facts.

* An increased command over the musketry-fire is pointed out in

of officers in a Prussian company is much less in proportion to the number of men than it is in our service, the number of non-commissioned officers is much greater. The conditions of a compulsory service give means for selecting an exceptionally intelligent body of the latter. Whether in our case it will be possible to work in the same manner is not quite so certain. It may be that experience will show that, taking all our circumstances into consideration, it is advisable to work rather with a very few men under the hands of an officer, and to place an intermediate link between the officers directly in charge of knots of men and the commander of the company. By some means or other authority must, it would seem, be made to pass upwards from the very lowest grades by a less rapid multiple; fewer men in each rank must be placed under the immediate orders of the next above them. As for many reasons tactical convenience rather suggests the use, if possible, of some one or two ranks of non-commissioned officers to effect this object, I venture to suggest the possibility of placing, say, six men under a corporal, and two corporals under a sergeant, and four sergeants under a subaltern, so as to bring a body of considerable size—about 58 men, all told—so directly under the hands of the latter, that he may be able in all ordinary conditions, whether of skirmishing or

the new drill instructions as one of the especial objects aimed at.—Was enthält.

otherwise, to carry out his captain's instructions with reference to it.* This arrangement, if the section † be permanently under the orders of the same subaltern, and he is made responsible for it, will have this further advantage: from the moment that an officer has learnt his work, he will become accustomed to the habit of working with a charge which is sufficiently large to give him an independent interest in it. The first six or seven years of service tend to fix the whole style of an officer's work afterwards. If the habit be once acquired of being never intrusted with authority in even a limited degree, and of leaning always on the mere dictation of others, it becomes extremely difficult in later life for any man to shake himself free from it, and either willingly to assume responsibility, or—for the two things almost universally run together—to delegate power. Yet, for the present condition of war, a readiness to

* It is scarcely possible to discuss these questions completely here without going more fully into the subject than is suitable at this part of the paper. The motives which dictate the exact form of the suggestions will appear in discussing the form of attack. I have also there discussed the form in which attack should be made if we must adhere to eight companies; but the practical difficulty of placing such a number under the command of a battalion commander, with present forms of fighting, is too great not to be noticed.

† I use the expression "section" because, as I propose that there should be four of these divisions in the company, and there are at present four sections in a company, the word conveniently suggests that idea. The difference, however, between a casually-formed section, of which either a non-commissioned officer or an officer may be in charge, and such a permanent division as I speak of, is sufficiently marked, and, for tactical purposes, very important.

assume responsibility if necessary, a knowledge when to assume it, and a capacity for guiding others without dictating to them, are, as we must believe if we listen to those who have seen recent fighting, more needful than all theoretical training, than all other practical experience.* The same principle of giving a definite sphere of duty to each man, and of making him responsible for it, applies strictly to the lower grades. As far as possible each corporal must be responsible for certain six definite men, and each sergeant always for the same twelve. Each should be as much understood to be so in his own degree as a captain is known to be responsible for his company. Nay, do not even our captains now, merely in consequence of an alteration in seniority, change the commands of their companies too frequently?

In deciding on the number of sections in a company, and of companies in a battalion, the following

* The following quotation from the Journal du Colonel Fay, extracted from 'Les Maréchaux de France,' though it does not relate to a tactical subject, shows so admirably how tactical *in*efficiency may be prepared for by training, that I give it as probably unique. It is a statement "des ordres adressés au général commandant la division de Metz le 7, *alors que les soldats vaincus de MacMahon passaient les Vosges dans une fuite désordonée. On l'invite à 'déléguer au commandant d'armes d'Epinal les pouvoirs nécessaires pour préparer et organiser la défense des passages des Vosges.'"—Fay, p. 54; Les Maréchaux, &c., p. 80. Whilst the Prussians were training their subalterns to work freely under assigned directions, the French had to free their marshals, in order to allow them in a specific exceptional instance to free . . . and so on to the lowest step of the hierarchy.

considerations are to be noted : Every body which may be required to work independently needs to have an advance to throw into the fight, special portions charged with the care of its flanks, and a reserve to enable the effective action of the body to be maintained as long as possible. For these reasons the multiple four, as we have it now for sections in a company, appears to be an extremely convenient one, both in that case and as to the number of companies in a battalion. This would also give companies between 200 and 250 strong (allowing for the additional staff necessary for a much-extended command), probably as large as could conveniently be under the direct personal superintendence of one officer. Battalions would then be nearly of the strength assigned at present to them by our new table for active service.

Another question arises as to the number of ranks. There are great objections to be made against altering the normal formation in a matter which habit has so engrained into our army, unless very strong cause can be shown for change. As May happily puts it, " To change that which has become customary and deeply rooted in an army is always a critical matter. When all old rules are suddenly altered, no one knows where to turn." The great object to be secured appears to be, the utmost possible adaptability to circumstances. The Prussians, though their experience in this campaign appears to have furnished them with few instances in which volleys were employed,* have introduced into their latest drill reforms, for the first time, the practice of firing four-rank volleys. The whole experience of the most effective volley-firing ever made, that of the Peninsula and Waterloo, tends to show that its special value consists in the moral effect produced by it. That, as has been pointed out by Captain Laymann,† depends mainly on the *sudden* destruction of large numbers. Hence the more effective on the spot it can be made, the better will the volley fired carry out the fundamental idea. Moreover, four ranks or three may be

---

* Boguslawski, p. 78 and 167 (Trans., p. 84 and 179), points out clearly that, since a volley has no effect *against skirmishers*, while during the time necessary to prepare a volley fearful loss is caused by the skirmishers, it is no longer applicable.

I must, however, again assert the extreme difficulty of being sure, in these cases, how much is a mere special opinion of the writers, and how much is a statement of observed fact. If it is opinion merely, the authorities which may be quoted on the other side are legion. The retention of the four-rank volleys in the new drill is perhaps not in direct contradiction of his theories; because Boguslawski, p. 171 (Trans., p. 183), in his final summing-up, appears to approve of their being employed in drill.

† The evidence against volleys certainly comes with all the greater strength from the Prussians, because they have devoted exceptional pains to retain their advantages, as is evident from a comparison of Prince Frederick Charles's pamphlet and that of Captain Laymann with Boguslawski. The distinction between "four ranks" closed up and our present "fours" ought, perhaps, to be pointed out to non-military readers. In the one case the front is diminished, the men being supposed to stand as close together as in two ranks. In "fours," on the contrary, the front is the same as with two ranks, the men standing at an increased distance apart. Provision is made in our drill-book for assuming the closer formation, but only for exceptional purposes.

sometimes convenient for the sake of diminishing the front presented to fire during movement, and in order to have more handy masses for passing over dangerous ground, in cases in which strict formation is necessary. One rank will constantly be the normal formation, since it will be employed for skirmishing. Against the habitual formation, three deep, in order to obtain a skirmishing rank in the rear, we have the strongest of possible evidence in the gradual and increasing tendency of the Prussians to employ formations which suggest that they only retain three ranks at all from a desire to introduce as few changes in forms as possible.* Moreover, it was adapted to a time when skirmishing was an accessory, not the decisive mode of fighting, as it is now. The two-rank formation admits of extremely easy adaptation to the necessity either for one rank or for four, and therefore it would seem to be by no means advisable to change it, however little the grounds for the original adoption any longer exist.

Yet another matter of organisation requires to be taken into account. The Prussians have adopted a principle in their corps arrangements, the soundness of which seems to me indisputable. They have for the two arms of cavalry and of artillery

* There is something startling about the pæan of joy with which the writer of the 'Was enthält' receives the orders for the abolition of the skirmishing-division system, considering the extent to which we in England have been disposed to think them the most important part of the Prussian system.

what may be called a double organisation.* Before discussing its advantages, it will, however, be necessary to allude to two or three matters in which the experience of the late war appears to have forced upon men, who have seen it *on both sides*, conclusions by no means identical with those hitherto commonly held in our own army. Boguslawski considers that one of the most difficult and most important operations of modern war is (Boguslawski, p. 94; Trans., p. 101) "the deployment of great masses" of artillery, "with unity of command" (Boguslawski, p. 95; Trans. p. 102). "Whilst the most unpractised eye would remark the systematic deployment of division and corps artillery on the part of the Germans, one could not fail to notice among the French an *absence of combination* on the part of their artillery in most of the actions." V. D., writing here, as it appears, of what had personally struck him as visibly decisive in the action of the Prussian artillery, says: "Dans les grands mouvements, elle a presque toujours été employée par masses d'une certaine de pièces, et dans les manœuvres tactiques par groupes de 24 pièces qui composaient une division à pied" (p. 238).†

Unfortunately, among us the utterly unpractical nature of our higher artillery organisation has led to

---

* The corps organisation requires to be studied in order to have its perfection in this respect appreciated.

† See also Boguslawski, p. 56; Trans., p. 60; and Prince Hohenlohe, p. 13, 14, 15.

a belief that the battery is the all-sufficient artillery "unit"—an expression which has perhaps gained currency not a little from its vagueness. Yet it is not difficult to account for the observed fact. There is scarcely another arm of the service whose special characteristics so demand the practice of combined action by large masses. The form of artillery concentration is always at its best when it consists in a concentration of fire and dispersion of pieces, however little this may be possible at all times.* Moreover, the effect of each gun is at its best when it is firing so as to take the troops at which it is directed at least partially in flank. Now a gun is always able to take at least *en écharpe* any troops but those immediately opposite to it. Hence, clearly, the effect of each battery will not be greatest when it is employed in supporting those troops which are attacking immediately to its front, or in opposing direct attack. Yet the whole position requires to be taken into account in the dispositions of the artillery. No portion can safely be ignored, simply because considerations as to the best isolated effect of the several batteries would lead to such neglect. Is it possible, then, to imagine conditions which render it more important, that while the utmost possible liberty is left to captains of batteries as to detail, they should be in the habit of working together under conditions assigned with

---

* As to which see Prince Hohenlohe, p. 15.

reference to the larger combinations of the battle-field ? *

On the other hand, every one who has watched the mutual relations of artillery and infantry on service appears to agree in this: You can never fully instil into the minds of the mass of infantry soldiers the fact that the protection which they receive from artillery is not in proportion to the nearness of the guns to them. Hence, when artillery has, in order to fire with effect, to fall back into a new position, the infantry are apt to think themselves abandoned. Since the guns absolutely must fall back, the question is, how best to prevent this most inconvenient discouragement of the infantry. Will it not be safest to trust to the effect of *camaraderie* and the habit of working together? † The notion that men who have been long associated with them, and whom they

---

* There is probably no practice which is a less satisfactory preparation for the present action of artillery in war than a well-dressed line of twenty-four or thirty guns at accurate intervals in action, irrespective of the ground. Some batteries will almost certainly uselessly waste their fire. It is inconceivable that the best position for each of thirty guns should occur exactly at each successive eighteen yards. Yet how can more elaborate work be practised if batteries are only exceptionally formed into higher units, and these are often of no administrative connection ?

† See, as to another reason, Prince Hohenlohe, p. 12 : "If, therefore, a close connection between the artillery and the other arms is not established by the regular organisation of the command in such a manner that" a proper transmission of orders and information "somehow naturally takes place, then mistakes will always arise which prevent the artillery from carrying out the specified object either in its full extent or at the right time."

know well will not desert them, is sure to operate powerfully. Moreover, if comparatively small bodies of the two arms are accustomed to exercise together, the after-talk over a field-day among the men will lead to a much more practically valuable acquaintance of each with the character of the weapon of the other, than any lectures or explanations given by officers would produce. The habit of seeing the artillery habitually take up those positions in which their fire is most effective, would tend to the same result. At present, when small bodies are together in garrisons, usually, on any few field-days there are in the year, a change or two of front is made by the whole force, artillery and infantry, in line; and the practice tends rather to confirm than to dissipate prejudices. But practice based on the experience acquired in larger annual manœuvres would lead to a very different result.

That the same general principle applies to the cavalry seems equally clear. Where cavalry are to be the eyes of the body, it is very important that they should fit properly into the sockets.* Cavalry that has to be constantly bringing back reports ought not to feel strange among the body in which it is to find the officers to whom it is to report. Practical experience has probably brought home my meaning to most of those who have been engaged on kindred duties. Yet no phenomenon was more strikingly

* There are some excellent remarks on this subject in 'Standing Camps.'

exhibited during the late campaigns than the advantage gained by the use of vast clouds of cavalry, accustomed to work well together. For the battlefield, should large masses ever again act together, cavalry may perhaps still be prepared by drill. But it was in the wider movements, where no drill but only habitual association of large numbers affected the result, that harmony of action was most apparent.

It is these almost contradictory necessities as to the two auxiliary arms which the Prussian corps organisation appears so happily to meet. Four batteries form a division. In each corps d'armée there are three field and one horse artillery division of three batteries, the whole being formed into a regiment. Thus 90 guns are associated under one definite command, and are sufficiently trained to be able to work together, as was shown by more than one incident of the war.* Yet one of these artillery divisions of four batteries is definitely attached to a specific infantry division, another to the other infantry division, and one horse-battery to the cavalry division. The remainder forms the corps artillery. The principle with reference to the cavalry is strictly analogous.

* The latest report which has reached us is, that two regiments per corps d'armée are to be formed. Probably, however, this only means that a decision some time since arrived at will practically be carried out—*i.e.*, the garrison artillery of each corps will be permanently separated from the field-artillery.

There are always a thousand and one details which go to the history of a campaign, which we never have sufficiently clearly presented to us, and for which we must make allowance. If analogy is worth anything —if we may at all read between the lines of the accounts we have received—it is at least as much to the *camaraderie* due to cricket, football, theatricals, all that throws men freely together, as to mere drill, that we ought to look for what is most valuable in this kind of organisation for tactical purposes, so far as the lower ranks are concerned. What is gained by the previous working together of the higher ranks, and to what the gain is due, has been so happily explained by Fezensac in a passage already referred to (see note on page 39), that it seems unnecessary to allude to it further.

Before leaving this part of the subject, a question has to be answered peculiar to our own army. In what way could the militia and volunteers, who form the bulk of our forces for defensive purposes, be best employed, should they ever be called upon to fight? I suppose that most of those who have faced the subject have attempted to work out various schemes by which advantage should be taken of that great quality, the extreme perfection in rifle-practice possessed by many volunteers in far higher degree than by soldiers. I confess that, having thought the question out as carefully as I can, I am unable, from a consideration of what all writers who have seen recent fighting unanimously urge, to come to any but one conclusion. It cannot be doubted that there are among the volunteer corps bodies in all essentials as highly disciplined as any men need be. The fact impressed itself upon all who had to do with most of those who undertook the trouble of attending *the whole* of the autumn manœuvres last year. There are among the volunteers some who appear not even to have arrived at an elementary conception of what the nature of discipline is. The first are, under present conditions of war, invaluable ; the latter are much worse than useless.

I see no way of selecting those who ought to be employed, and of getting rid of those who would do mischief, except that of intrusting to each corps d'armée commander, at the moment when the services of the volunteers of his district are required, the duty of assigning their proper functions to each. Some will be fit to join any troops of the line, and to become the sharpshooters selected on each occasion, or the mounted riflemen, who become the eyes of the army. Others may be able to act if properly incorporated with good troops, as the Dutch Belgians were incorporated in Wellington's army of 1815. Others will only be fit to be thrown into a fortress, there to learn "discipline and drill."* In any case, for troops

---

* I would ask any one who doubts these conclusions to study Captain S. Flood Page's Lectures. I think, also, that a study of Sir Thomas Acland's pamphlet leads to the same conclusion. The only objection

without discipline there is no place in modern open war. Of the militia, *mutatis mutandis*, almost exactly analogous expressions must at present be employed.

to be made to what is urged by these and by other officers whose well-seconded exertions have made their corps what they are, is this : Sir Thomas Acland has pointed out that legally a volunteer corps depends for its existence on its commanding officer. Its value and efficiency usually depend on the exertions either of him or of his adjutant quite as absolutely. Now commanding officers or adjutants of creative faculty do not grow on hedgerows.

As regards the method of preparing the volunteers for the work they can best do, surely it is unnecessary to impose on them any uniformity of drill or of training which does not take into account the local circumstances of each corps. If drill is now rather, for the army, a means of securing effective discipline than a direct preparation for war, must not its applicability to the volunteers depend on the extent to which it does tend to bring them under discipline? Now it may surely be questioned whether, when the *only* training in discipline which men receive is that of following out prescribed forms, which they are bound to know as perfectly as their drilling officer, this result is quite secured. For instance, does it tend to teach them the very thing which it is most essential that they should learn—obedience to a man whom they fancy has made a mistake? The drilling officer makes a slip in the form prescribed. The men don't move. They are told they have done quite right. That is their only experience of discipline. What must be the effect of it? When mechanism was the mode of securing military movement, this was essential, for the machine could only move in one way. Is it so any longer? Nothing perhaps illustrates the change which has in fact taken place more perfectly than the difference between the notice which is prefixed severally to our own drill-books and to the Prussian. In the one, it is pointed out to commanding officers, that her Majesty having approved the regulations, not the slightest deviation from them is to be permitted. In the Prussian, the Emperor draws attention to the fact, that though he has approved the regulations, it is expressly to be understood that latitude to commanding officers as to modification of forms is no way withheld. Surely the volunteers may be content with a uniformity which is sufficient for the Prussian army.

ATTACK OR DEFENCE.

I HAVE noticed throughout the earlier part of this paper that the changes which are now necessary are required for offensive rather than for defensive action. I have taken for granted that the fact that those changes are needed in order to enable an army to assume the offensive, affords sufficient ground for asserting that they are imperatively required in every army. Troops cannot take the field which are not able to attack an enemy's positions. Nevertheless it is important to determine whether, on the whole, an army acts under the most favourable conditions when it is defending a position or assailing one. It is necessary, also, to make clear the general relationship established by modern circumstances between offence and defence on the large scale and in detail. Very many points of great tactical importance depend on the solution of this problem. In treating each of the four minor topics of the present essay, it will be necessary to make some assumption in regard to it.

The difficulty of arriving at a definite conclusion consists in the tendency of the subject, if one may venture on the expression, to turn round in one's hand as one deals with it. I can hardly make this clear better than by referring to a question on the borderland of tactics, which may conveniently here be noticed, because it happily introduces what I may call the first turn of the tactical problem.

Shortly before the commencement of the war of 1870, a discussion took place at the United Service Institution on "Offence and Defence." A most able lecture had been delivered, intended to point out how great an advantage the defence now possesses in the matter of weapons and of fortification. A gentleman not a soldier was present, who often supplies to the discussions at the Institution that most useful kind of information,—hints as to the general impression left by the lectures on the non-professional public. The deduction which he made was this: It was a criminal thing for France to maintain an army so large as she then possessed, because 100,000 men were sufficient for the defence of any country; and 500,000 could only be required for invasion of other lands. The Institution papers had hardly reached the members before this so enormously too great army had been swept away by the tide of invasion. Time sufficient had, however, elapsed to allow of the lecture being translated into the 'Revue des Cours Scientifiques,' where, for the encouragement of Paris, it, with this wonderful statement, appeared two days after Gravelotte.

But in what way, in a strictly tactical sense, was it that the most brilliant successes of the invaders were achieved? Surely by employing to the full the advantages now possessed by an army which can force another to make direct attacks upon it. In the first period of the war, undoubtedly, the Germans were successful in a series of minor offensive engagements of very great importance. But in all but one of these they completely outnumbered their opponents; in that one case of Spicheren, during the early part of the day, the Germans were utterly unsuccessful; and by night, when the position fell, they were in numerical superiority. In each of these cases the loss of the victors so enormously exceeded that of the vanquished, that it would have been impossible for any army to have long borne the exhaustion produced by such work.* No doubt, as has been admirably shown by Captain Laymann, "loss" is a most inadequate test of the results of battle. Our object is to make men run rather than to kill them — to break the effective force of the army opposed to us by destroying its moral power and cohesion, rather than by placing a certain number of men *hors de combat*. The Prussian loss was well worth incurring. If two or three army corps acting together can bring to battle a single isolated corps, they ought, no doubt, at all

* Streffleur, Dec. (quoted more at full on page following).

Attack or Defence.

hazards to do so. It is almost as certain now as it was formerly, that the destruction to the effective power of the attacked will be so great as to far more than compensate any losses that may be incurred in the attack. But surely from these earlier battles one would not infer that on equal terms the offence ought to have the advantage.

At Rezonville, again, that which was most remarkable was the tenacity with which Alvensleben, against tremendous odds, held, till supported, the ground he had won by surprise from troops on the march.

What is the lesson on this subject of the battle of Gravelotte? Soldiers who had fought for hours with marvellous success under the most unfavourable circumstances against overpowering numbers, were, by the fault of some one or other, left in the very key of the position without any ammunition.* All agree

---

* The exact facts on this subject are so important that I quote here at full Captain Brackenbury's account of the circumstances. It should be remembered that, as may be seen from the note on page 29, the corps had very many young soldiers in its ranks. It had only reached its position after nightfall, having been left on the 17th at Verneville, alone of all the corps on the further side of the great ravine which separates the Gravelotte slopes fron those of St Privat, Point du Jour, &c. It is curious that this fact seems to have been entirely unknown to the Germans. Borbstaedt expressly states that no change took place in the dispositions made on the first retreat from Rezonville:—

P. 133.—"Le corps Canrobert, qui fut placé par Bazaine sur ce flanc et qui occupa Saint-Privat en remplacement du corps Ladmirault qui se rapprochait de sa gauche, était de tous les corps celui qui avait l'artillerie la moins nombreuse, et n'avait point de génie ni d'outillage pour les travaux de retranchement. Conséquemment, lorsque l'attaque des Allemands se prononça contre le flanc droit, elle trouva les troupes du 6ᵉ corps, en

that the special weakness of the French was a tendency to waste ammunition, and to neglect the preparations necessary for providing fresh supplies. It could not therefore be inferred from the action at St Privat that men on the defensive are specially apt to be left without ammunition. It is in fact notorious

rase campagne, absolument exposées au feu de son artillerie, et avec une artillerie incapable de répondre d'un manière efficace. Ce corps n'était pas faible en artillerie seulement. Il avait éprouvé le 16 des pertes plus considérables qu'aucun des autres corps. Il avait dépensé la majeure partie de ses munitions ; pour gagner le 17 sa position, il avait marché plus longtemps que les autres, et, par suite de son mouvement, n'avait pas reçu de nouveaux approvisionnements de munitions. De telle sorte que la partie de la position qui demandait la plus vigoureuse résistance était dévolue au corps le plus affaibli. On n'envoya même pas à son secours, ni artillerie de réserve ou de la Garde, ni un homme ni un outil du génie de la réserve pour donner les moyens de mettre les soldats à l'abri des projectiles allemands."—P. 134.

P. 140.—"Jusqu'à 4 heures, la gauche du corps Canrobert n'avait essuyé que le feu très-accablant de l'artillerie de la Garde. Peu de temps avant il dut abandonner à l'infanterie de la Garde l'avant-poste très-exposé de Sainte-Marie-aux-Chênes. Les munitions d'artillerie de ce corps étaient déjà presque épuisées. Les 84 bouches à feu de la Garde prussienne purent donc se rapprocher de la position française, tandis que 36 canons de l'artillerie de réserve saxonne, plus 48 pièces appartenant aux divisions du corps saxon, qui étaient entrées en ligne entre Sainte-Marie et Jœuf, venaient s'ajouter à celles-ci.

"Le 6ᵉ corps se trouvait ainsi attaqué par non moins de 168 pièces dont 120 dirigeaient leur feu sur Saint-Privat ; pendant que d'un autre côté les munitions étaient tellement épuisées qu'il etait nécessaire de ne jeter qu'un projectile par quart d'heure à chacune des 76 pièces, afin de réserve quelques coups de feu pour les éventualités qui devaient evidemment se présenter.

"Canrobert envoya emprunter des munitions à Ladmirault, et *les deux seuls wagons que ce général put lui accorder étaient en route lorsque la Garde prussienne attaqua Saint-Privat.* Trois brigades avancèrent en ligne de colonnes, précédées de tirailleurs de Habonville et Sainte-

that, under ordinary circumstances, the defensive force ought to have a great advantage in the facility with which fresh supplies can be provided for the fighting line. Till ammunition was exhausted no force seems to have been sufficient for the attack. Finally, panic spread among troops whose sole means of defence

Marie ; leur attaque était couverte par le feu concentré de leur artillerie. Leur front s'étendait sur 2000 pas ; ' Mais l'effet du feu de l'ennemi fut tellement écrasant que, d'après les rapports, près de 6000 hommes succombèrent dans l'espace de dix minutes, et l'attaque dut immédiatement être discontinuée.'—(Duke of Wurtemberg.) *Ce résultat est attribuable entièrement au feu de l'infanterie française. Mais vers 6 heures les munitions manquaient autant à l'infanterie qu'à l'artillerie.* Les gibernes des morts et des blessés avaient été mises à contribution par les survivants ; mais il ne restait pas aux soldats une provision suffisante pour opposer un feu nourri dans le cas d'une seconde attaque."—P. 141.

In which it is to be noted that the first great attack of the Guards was repulsed by infantry-fire alone—that that infantry, unsheltered by any intrenchments, had been exposed for hours to perhaps the most tremendous artillery-fire ever directed against one part of a battle-field. Yet it was under these conditions that that loss was inflicted, which has made solid columns of attack against modern weapons a byword for all time to come. It is rather remarkable that with this positive testimony to the fact that it was impossible for the sixth corps to have fortified its position at all, we should have on the opposite side the assertion made by Borbstaedt, p. 335, 360 (Translation, p. 434, 461), that the position at St Privat had been strongly fortified by the French, with their accustomed skill in such matters. It is of course possible that St Privat had been hastily prepared for defence by Ladmirault's corps, before it abandoned the position, otherwise one is disposed to think that the defences must have been only such loopholes as could be made with whatever came to hand. In any case, stone houses are an odd protection against artillery-fire. In other respects St Privat was doubtless a magnificent position, notably in the glacis slope to its front, and in the perfect protection afforded near at hand to the supports. The Prussians seem to have suffered seriously in the final attack, despite the feebleness of a fire which depended on the stray cartridges of the fallen.

was gone. The Prussians and Saxons entered St Privat, and the remainder of the position became untenable.

Granted that the forms of attack were on neither side adapted to modern conditions. What I desire to bring out here is merely that there is not as to this part of the war that necessity for accounting for the successes of the assailants, which naturally seemed to the Duke of Wurtemberg to exist when he received the telegrams. "The Bavarians have taken Weissenburg at the point of the bayonet, and the Prussians have carried the Geisberg in their first assault."* When, in fact, fully prepossessed by this idea, he started off to study the method which had enabled the Prussians to establish the supremacy of the attack over the defence.

Turning now to the really startling phenomena, what was it which made the victory of Sedan so complete? Was it not the utter incapacity of the French, trapped in a net, to break through in any direction? As so often before, the German attack was in the nature of a surprise. The advantage which it secured was that it left the French the choice between surrender and *attacking* under impossible conditions. Surely the sieges of Paris and of Metz, which I have discussed at the beginning of this paper, present the same feature of strangely successful defensive tactics.

* Duke of Wurtemberg, p. 5, 6.

May not the facts then, on the whole, be summed up thus ? In those cases in which the German victories were obtained under conditions such as in any conceivable state of warfare would insure success, there they were due to bold attacks, in the course of which the victors suffered fearful losses, and till the moment of victory inflicted hardly any.* In those cases in which the German victories were strange, unprecedented, overpowering, they were due to a brilliant application of offensive strategy, but of defensive tactics. That is to say, that during this latter and more successful time, what the Germans did was to take up positions in which the enemy's army was forced, for the sake of other considerations than those of merely tactical advantage, to assume tactically the offensive on the grand scale. In the cases of Metz and of Paris, the considerations which forced their enemies to attack were as distinctly those of supply as if the Germans had in the ordinary strategical sense threatened lines of communication. In the case of Sedan, the phenomena would inevitably have been the same had not most peculiar topographical conditions rendered the situation more immediately intolerable by reason of the effect of the German artillery.

It is, however, most important to state quite clearly how far this brings us in the inquiry. It appears to be established, by a comparison of these facts with

---

* See especially the article in the 'Militarische Zeitschrift' for December, referred to in the note to p. 82 of this book.

those of the earlier portion of the campaign, that it is of very great advantage to an army to be able to force another to attack it. Moreover, in the particular cases noted, this result appears to have been due almost directly to the defensive power of the weapons. But it would not necessarily directly follow that because it is of advantage to an army to fight a defensive battle, that therefore weapons on the defensive have a very great superiority, nor even that small bodies on the defensive have an advantage over those assailing them. I have noticed that the superiority of armies tactically on the defensive tends to favour bold offensive strategy, and therefore invasion. It may be, similarly, that the mere local superiority of small bodies on the defensive, if it exists, does not only and always tend to the advantage of defence on the grand scale. It may sometimes favour in battle bold tactical offence, which seizes at some loss positions which the army originally on the defensive must attack in order to avoid being compromised. I believe that this is in fact true, and that in general the real ultimate advantage to the defensive army will be due to more complicated causes, to be noted hereafter. It is clear, however, that between such positions as that held by the French at Gravelotte and those of the German armies awaiting attack round Paris and Metz, there is in this respect a marked distinction. Throughout the day at Gravelotte a series of utterly unsuccessful and frightfully costly local attacks were made, first on one side and then on the other, along many parts of the position. Now, clearly, the greater the power of resisting local attacks along the line possessed by small portions of their own forces, the more possible was it for the German leaders to accumulate, without danger and unknown to the French, overpowering forces against St Privat. Moreover, the French on the defensive at Gravelotte incurred the risk of having the advantages of the defensive turned against them, by being surrounded and forced to attack. The Germans surrounding Paris or Metz, on the other hand, would reap in almost any event the full advantages of a local superiority of the defensive. The fact, however, of this local superiority of the defensive, is itself disputed on theoretical grounds, and must be independently investigated. The advantage of defensive tactics has been also doubted. I propose, therefore, to consider in succession the reasons on account of which various recent writers* deny the practical superiority of defensive tactics, and then to examine the minor question of the local superiority of offence or defence.

* Captain Laymann throughout; Duke of Wurtemberg, p. 40; German General, p. 247; Captain May, Tact. Retro., p. 41. See also Streffleur, October 1871, maintaining the same.

On the other hand, Boguslawski decides on the other side, p. 144; Trans., p. 155, "Attack of an inferior force against a superior almost hopeless."

The Duke of Wurtemberg, however, p. 39, finally appears to admit that it is an advantage to prepare for offence by the defensive, and with this Das heutige Gefecht, p. 5.

"It is all very well," Captain May urges, "to talk of the advantages of defence, if your enemy will only come and attack you. But will he?"* The argument, in fact, surrenders the whole case; for no one would be so mad as to maintain that an army must not be ready to strike offensive blows. The question rather is, ought not an army to be constantly watching for opportunities of striking, *under circumstances of special advantage*, offensive blows against weak portions of the enemy, in such a manner as to oblige him to attack on a larger scale, and therefore with greater loss? That such chances will still present themselves is surely a lesson fairly to be deduced from the late campaigns, if we eliminate from our examination of them circumstances irrelevant to this question. If it be urged that these opportunities were afforded only by the defective nature of the French organisation and strategy—that such errors are never likely again to be committed—in the first place, this reply may be offered: Mankind has for a long time been engaged in the study and practice of war; yet on one side, at all events, if not on both, errors have always been committed in every recorded campaign. But further, strategical considerations have not become *less* important now that the supplies of ammunition which an army requires are much more bulky than they were. Lines of supply have not become less vulner-

* Tactical Retrospect, p. 41: "If only the enemy will do us this pleasure."

able now that they depend so largely on railways and telegraphs which a few men in a short time can render unavailable for days.* If it be maintained that it is always at least possible for able generals to avoid the necessity of attacks, it may be asked, By what possibility could the Prussians have avoided attacking an army which had established itself in a position directly threatening the lines of supply of the besiegers of Paris, if the threatening army had at the same time properly covered its own communications with the south? Such a position might certainly have been assumed during a very considerable portion of the siege of Paris. That the Germans were ready to incur the risk, showed a contempt for raw levies which was fully justified by the event. The broad inference which is suggested by a study of the facts is not affected by that circumstance.

"Whatsoever your theoretical arguments may be—in practice," urges Captain Laymann, "the offence nearly always wins."† I remember once to have heard a boy ask, "Why is it that people take the trouble of building fortresses, since those inside seem to be more often in the end beaten than those outside?" The

---

* Von Moltke, p. 10: "A skilful general will *often* be able to choose defensive positions of a nature so strategically offensive that the enemy *will be obliged* to attack them." Had it not been disputed, one would have thought it rather calling on a sledge-hammer to crack nuts to invoke Von Moltke's authority on such a subject.

† Laymann, ch. ii., "Attack or Defence."

chances, he seemed to think, would have been about even, if only the two bodies could have met fairly in the open field; but the stupid plan of building a fortress had turned the scale against those who were in it. The boy was not aware that a fortress is a means for enabling a weak body to resist, at least for a time, the efforts of one far more powerful to crush it. Such surely is also the case as to this question of the practical result between attack and defence. The offence has no doubt in the majority of cases been successful, because, as a rule, the assailants have adopted their *rôle* in consequence of their superiority either in moral or material power. One instance of successful defence ought therefore to outweigh many on the other side. As a matter of fact, the history of our own fighting in Europe, at all events under our greatest chief, presents a majority of victories for the defensive. This deduction from the past is clearly a fair one in reply to Captain Laymann. For essentially his assertion is that the experience of the past shows that, no matter what changes may take place in armament, we are liable, from theoretical considerations, to undervalue the moral advantages of the attack. In all that he urges as to these there is surely immense truth. But that the English army of old time was not taught to undervalue them, let Marshal Bugeaud testify.* It was a defence prepar-

---

* See the famous description quoted by Trochu of Bugeaud's experience of fights with English troops, Trochu, p. 239 to 243.

atory to attack which formerly paved the way for victories. It was the perfect adaptation in defence and attack of our mode of warfare to the characteristics of our soldiers which won them.

Passing from these more or less philosophical considerations, we are met by arguments deduced from the relative advantages of the action of weapons on the offence and on the defence respectively. The Duke of Wurtemberg* maintains that in one respect, at all events, the nature of the new infantry arm has made it available for offence, whereas formerly the firearm was essentially the weapon of defence. This is no doubt true, so far as it goes. It involves a complete change in the whole system of offensive tactics. But that the offence has recovered on this account what it has lost in other ways is by no means so apparent. According to the Duke of Wurtemberg himself, and according to most recent writers, the great change which has come over the face of fighting is this : † Formerly all firearms prepared the way for the bayonet. Now the bayonet, or rather the charge, in which the bayonet counts almost for nothing, is only the means by which the fruits of the fire-action

---

* Duke of Wurtemberg, p. 40.

† Except that Boguslawski, p. 167, Trans., p. 179, seems to think that even a bayonet-charge has more chance than a *volley* under some circumstances : no such idea as a regular charge, in the old sense, enters into any of the descriptions of battles of Boguslawski, German General, or the Duke of Wurtemberg at all (see latter, p. 39, especially).

are reaped when they are fully ripe. Formerly the firearms were used to induce such a state of things as would make it possible to bring the bayonet to bear. The fact of a bayonet-charge then implied that the critical moment had come. Now the rush to seize a position implies that the critical moment *has passed*, or the rush is sure to be fruitless. The reason why of two bodies, each employing the bayonet, the assailant had the advantage, is apparent at once if it be remembered how almost entirely moral the effect of the weapon was. Hardly ever did any troops wait to receive the attack of bayonets actually carried steadily at them. It was enough to secure the opportunity for delivering the charge. Hence the ultimate arbiter of battles was formerly a weapon in the use of which the advantage lay wholly with the offence. The weapon which was essentially the instrument of defence now decides the issue. True, it has become partially available for assailants during their advance. Does it follow that the result is to make it *more* effective in the hands of assailants than in those of defenders? As regards infantry *fire*, the inconvenience from which the assailant suffers is this: A certain amount of time must be spent in moving forward, and during that time the men advancing cannot fire, and are more exposed than the defenders. Of course this may be compensated to some extent, as it is in the new system, by hastening the pace, and by keeping up the fire of certain men whilst others are moving forward. In this respect the attack has gained something by the change in the arm, since the rapidity of loading renders it possible to employ the weapon for offence in the hands of all soldiers, not merely in those of picked corps.* But nevertheless there must be a certain loss of fire during the period actually spent in the advance. It is inconceivable that even the aim of those assailants, who at each moment have reached a spot from which they can fire, should be as steady as that from a fixed position. Now, this advantage of the defence is obviously in proportion to the distance over which fire is effective, and the number of rounds which can be fired within the time which is required to traverse that distance. Of late years, both the distance of effective fire and the number of rounds per minute have greatly increased. It is true that when the assailants reach the position and succeed in penetrating a gap in it, the defence of the adjacent portions would collapse if the defence depended on single lines without power of reinforcement. But in fact the defensive position never is so formed, and the defenders can move some of his troops whenever necessary to take the assailants in flank. Hence in these respects the two bodies are perhaps ultimately on an equality, but only after the assailant has been exposed to losses great in proportion to the charac-

* This is the special advantage which the Duke of Wurtemberg claims for it.

teristics of the weapon already referred to. Hence, merely considering the relationship of one body of infantry on the defensive to another assailing it, it needs little evidence to convince us that if the fight is now decided by fire, not by the bayonet, the defensive body must be able to repel the assault of very superior numbers. In fact, this is confirmed by all the evidence we possess. Wherever direct assault has been successful, it appears to have been due to the prior action of artillery.

It becomes therefore exceedingly important to inquire in what relation of relative advantage the artillery of offence and the artillery of defence stand to one another. The German General* urges that the artillery on the offensive has this advantage over the artillery of defence, that the former is firing at a stationary, the latter at a moving object. Now no doubt as soon as the artillery on the defensive is able to fire at the assailing infantry, it has to fire at a moving object; while as soon as the artillery on the offensive can find out where the infantry on the defensive is, it has to fire at a stationary one. But there appears to be one very important element in the calculation which the General has ignored, for reasons not difficult to discover. The enormous practical superiority of the German artillery over the French rendered any duel which occurred between the two arms, under almost all circumstances, favourable to

* German General, p. 238, &c.

the Prussian, which was usually the assailing artillery.*
But according to all showing, the first duty of the
defensive artillery is to protect their own infantry
from the hostile artillery by opening fire upon the
latter. If, whilst the attacking artillery is wholly or
mainly employed in pouring fire upon the defensive
infantry, the defensive artillery is able, almost without being fired into, to search out positions occupied
by the guns, in what respect can it fairly be said that
the artillery of offence has an advantage over the
artillery of defence? " A fight where you beat and
I am only beaten," used not to be considered the
description of contest in which he who sustained
it had a great advantage surely.† Yet such is the
kind of action which is forced on offensive artillery
by the necessity for concentrating all possible means
of destruction upon the defensive infantry. Nor does
the offensive infantry gain in facility of approach

---

* It is true that the General does himself point out that the Prussian artillery has been most successful in offence, but this hardly meets my point.

† It is extremely important to note the change which has here occurred in the relationships of the two artilleries. Captain May was the first to point out, that now that guns are not effective at canister range, as all admit, and that infantry is so terribly powerful, "to silence the attacking guns is the essential object of the artillery on the defensive—then the infantry will know how to repulse the enemy's infantry; but artillery on the offensive should, on the contrary, make it their principal object to play upon the infantry of the enemy" (p. 51). This result, due to the immense *defensive* power of the infantry arm, must leave a balance of advantage to the artillery of defence surely, and therefore again diminish the effect of the other on the infantry.

because the fire of the artillery is withdrawn from it during this time. Till the infantry on the defence has been crushed, those on the offensive cannot approach at all. It is surely very hard to believe that if the defensive artillery of each part of a position concentrates its fire in succession upon gun after gun, the fire of the offensive artillery will possess altogether that superiority in the efficiency of its practice which has been claimed for it. Great as the advantages of firing at a fixed object may be, and convenient as it may be to be able to have a somewhat freer *choice* of positions, the extreme advantage to artillery of being able to shift its position very often is by no means apparent. The facts in this respect have been so happily brought out by Sir G. Wolseley, that I can scarcely do better than quote them as they stand. "Formerly, when guns were taken up to within a few hundred yards of the enemy's position, it was easy to lay them with precision for such short distance; but to do so at long ranges is a different matter, requiring time and very great nicety, as the exact distance has to be ascertained. The fewer the movements executed by a battery, the longer it will be in a position to inflict damages upon the enemy; for it is a recognised axiom, that guns are useless when limbered up. It is therefore of great consequence that good positions should be found for the artillery before the action begins, and that when posted there it should not be moved, unless the enemy succeeds in bringing a musketry-fire to bear upon it, or that in being driven back he retreats beyond its range. A few hundred yards either backwards or forward make but little difference in the effectiveness of fire from rifled guns."* There is surely no reason why a very considerable latitude of movement should not be possible. Indeed practically, on most battle-fields of the past, the difference in this respect has not been considerable. Yet again, all recent experience has shown, not only that it is essential that the way for infantry attack should be prepared by heavy artillery-fire, but that the operation of preparing it is a very long one, if the defending infantry are good troops. The longer the operation lasts before the assailants are able to move to the attack, the more must their artillery suffer from a fire to which it is not its business directly to reply. Will it be possible, under such circumstances, to prevent an artillery duel from preceding the infantry attack? If it does, is it evident that under fairly equal conditions the artillery of the defence ought to be beaten? An army on the defensive has usually had time to get into position a more powerful armament, in proportion to numerical strength, than its assailant. Even if calculations based on the advantage to the defender of the superior knowledge

---

* Sir G. Wolseley, p. 252. No one has, however, brought out more forcibly than the German General himself the necessity, under conditions of long-range fire, for careful aiming, and therefore, it must be presumed, for few changes (p. 238, 239). Prince Hohenlohe (p. 37) considers that, as a rule, guns should not move forward less than 1000 yards at a time.

of the ground which he ought to possess are apt, as Captain Laymann* says, to be erroneous, at least with our present range-finders, the artillery in defence must have been very improperly prepared for their work, if they do not know the distance and general nature of the ground of every important position which the assailant can take up. The assailant can scarcely obtain an equal degree of knowledge of the whole of a position properly occupied.

Therefore it seems to me, despite the weight of authority on the other side, impossible not to infer that the defence has gained enormously by the new weapons, in so far as the expression "the defence" is intended to refer to the direct attack of one body of troops and the local result at the time.†

On the other hand, there are some respects in which this local superiority of the defence will very frequently in practice tell in favour of attack on the large scale. For now, as always, the assailant will

* Laymann, p. 13.

† The question seems to me to be practically settled by an article in the 'Militarische Zeitschrift' for December 1871. The writer has no intention of arguing on one side or the other, but he puts in a series of pages the losses in every battle (p. 200 to 208). The general result is this: roughly speaking, where no special cause interfered, if the French were on the defensive, their loss was about *half* that of the Germans; if the French were on the *offensive*, their losses were as ten to three. My argument in the text is, that our object ought to be to face the loss of two to one, if necessary, where relatively small numbers are engaged on both sides, in order to be able to inflict the loss of two to one when large numbers are engaged. It was this which, as Blume shows, the Prussians carried out with such success in their campaigns in the east of France.

possess, at the commencement, at all events, of an action, the incalculable advantage of the initiative. He will therefore, to a great extent, be able now, as formerly, to determine, unknown to the defendant, at what point of the battle-field the decisive crisis shall take place. Now also, as always, his object will be to deliver his blows at that point with all the surplus force which he can spare from the remaining portions of the battle-field. Since, then, his object will now, as formerly, be to maintain a simple equilibrium for as long a time as possible throughout the rest of the field, it will be a great gain that his line can be secured by fewer men, and that it will be less easy for the defender to discover any weakness at those points which have been somewhat denuded. The vast accumulation of *masses* at particular points will be no longer the road to victory, as it so often was formerly. The object will be to weaken that portion of a position against which assault is intended, by drawing off the enemy's force to other points. This will often be accomplished by considerable extension towards the flanks.* The facilities for so extending without danger will be greatly increased by the enhanced defensive power of comparatively weak bodies.† In general,

---

* It is on this necessity for being able to extend in attacking that Boguslawski bases his belief that attack can only now be made by very superior forces, p. 144; Translation, p. 155.

† Though whether it will now be possible for the assailant to remain throughout the greater portion of his line on the defensive may be questioned. This point will be discussed by-and-by.

if both extend against one another, the army ought to have the advantage which, having prepared for attack, is awaiting the opportunity of striking at weakened points when the extension has reached its limit. In many cases, also, the local defensive power may enable a general to repeat the operation which the Prussians so habitually practised throughout the late campaign, that of continuing the extension till the whole opposing army has been actually outflanked, or even surrounded.* Thus the very defensive power of the new small-arm may often tend, by giving greater freedom to the attack, on the whole to confer advantage on the assailant.†

But—and here again the question turns round under consideration—it is by no means evident that if the general who is at first on the defensive appreciates strictly, at its right value and no more, the defensive strength of each part of his own position, and is ready to assume the offensive when opportunity offers, he will not be able to meet this kind of action with every prospect of success. In practice it will be extremely difficult to combine the necessity for a long previous preparation by artillery-fire, with an adequate con-

---

* The German General, Boguslawski, and the Duke of Wurtemberg, seem habitually to assume that the only attack of the future is by this extension actually round a flank; but when this is applied to the extreme flank of the whole position, surely it is driving the point too far. Both armies will extend; the difficulty is to judge correctly how far to extend in each case. This has been further noticed in discussing the defence.

† Boguslawski notes this, p. 63; Translation, p. 68.

cealment of intention as to the part against which the weight of the attack is to be thrown. Usually, therefore, the advantage, even in preparing for ultimate attack, ought to rest with the army which, without committing itself to any preconceived scheme, but perfectly prepared for all emergencies, awaits the movements of its adversary.

Hence, on the whole, I must in all humility maintain that the arguments adduced are not conclusive against the advantages of the time-honoured habit of the English army, that of awaiting attack in order to return it. Nor, as I have already urged, has it ceased to be possible to force an unwilling adversary to accept the risk in which it involves him. Looking at the question from a specially English point of view too, there are serious reasons for thinking that the tactics which, during the last generation, were specially adapted to our national characteristics, are in their large and general features still those which it is best for us to employ. We, better than other nations, can afford to play a waiting game—can less afford to break our strength against strong positions. Of all nations, the Prussian is the one which, in all these respects, is least of a model for us. Prussia, with her magnificent organisation and her capacity for bringing at once into the field the whole force of the nation, did well to make her unprepared adversary pay for his unreadiness, immediately and at any sacrifice. The naturally poorest of the nations of Europe, she could less than any wait with her ground untilled, and her army costing, despite all her economy, day after day enormous sums. For us, whose military strength at the outbreak of war never does, and never will, represent a fraction of the force which the nation, when once fairly engaged, is ready, if it has the time, to throw into the arena, delay is almost certain to be gain. To us, therefore, the taunt, "Yes, if only your enemy will attack you," is exceptionally inapplicable.

But far more important than this is it that the characteristics of our soldiers are essentially good for defensive rather than for offensive struggles. According to all showing, the "preparation for attack" consists far more in shattering the nerves of the men who finally hold a position at the moment of assault, than in diminishing their numbers. There is a certain something which has made itself known in these last times of war, often noticeable, never quite definable,—often hinted at, never fully recorded,—a certain shock, one scarcely knows whether to call it to the nerves or to the spirits of men—probably to both combined. It seems to answer somewhat to the old cannon-fever, but to be an altogether new thing in its intensity, and in the overpowering decisiveness of its effects upon the capacity of men for action. It appears to be largely due to the suddenness of the loss inflicted by infantry-fire, as much as by the general crash of artillery. Those who, knowing war, read the accounts of some of the actions round Metz, are convinced that in more than one instance this, which seems to be an almost involuntary physical illness, came over the French. There are certainly remarks in some of the reports which seem to suggest it. Now, if there are any troops in Europe who, from their previous history, are likely, unless their characteristics have utterly changed, "not to know" when a position has been in this sense properly "prepared," those troops are ours. The form in which this old British ignorance may now best display itself is much changed. It may be a fatal, not a useful quality, if it is accompanied with some other forms of ignorance, which it will take much labour to eradicate from our army. But that it is a quality inestimably valuable when it leaves men, as it used to leave our soldiers, ready, after being pounded for hours without possibility of reply, to fall back in order from a position, or to advance to an attack, cannot be doubted. Unfortunately we have reason to be aware of a capacity in the British soldier, when the bonds of discipline have been unduly relaxed, for acting in quite other ways. The care which is needed in the future to keep men, not less absolutely because they are less tightly, in hand, needs, for the preservation of this old characteristic, if for nothing else, very anxiously to be considered. Our soldiers are not at this moment, nor are they likely for many a long year to become, the most highly educated in Europe. Let us by all means do our utmost to remedy this defect. It is a preposterous libel to pretend that it is an ineradicable one. The fact is due to an old system, for which the previous training was admirably fitted. That system can exist no longer; but till we have fairly shaken off its trammels, let us prefer that form of fighting which gives us the best opportunity for the use of those high qualities which under it were developed. It may be that our soldiers will not at once recover as an army the reputation which Colonel Gawler justly boasts that they always possessed as skirmishers, where they were known by experience, not by vague report. But, at least, unless all that was taught by our grand old discipline has disappeared, they will not skirmish much worse after many an hour's heavy artillery-fire than they would have done before it. It has always been when both sides have been long pounding that the true worth of our soldiery has come out.* This is the precise trial to which modern war subjects infantry on the defensive. If they can meet it, and are not utterly overmatched in numbers, victory is theirs.

---

* "La mort était devant eux," says Foy of our infantry at Waterloo; "la honte derrière. En cette terrible occurrence, les boulets de la garde impériale, lancés à brûle-pourpoint, et la cavalerie de France victorieuse ne purent pas entamer l'immobile infanterie britannique. On eut été tenté de croire qu'elle avait pris racine dans la terre, si ses bataillons ne se fussent ébranlés majestueusement quelques minutes après le coucher du soleil, alors."—Foy, p. 323.

RETAINING POWER OF SMALL BODIES.

FROM a general consideration of the facts I have been urging, another very important deduction may also be drawn. The temporary retaining power of a relatively small number of men has very largely increased, and this is yet more notably so in the case of small bodies expecting reinforcement. If infantry has acquired so great a power of resisting the attacks of infantry or cavalry, it results that the only method by which a position at all favourable to defence can be carried, is either by first accumulating against some special part of it an overwhelming fire of artillery, or by long turning movements. Now a body expecting large reinforcements is to a great extent protected against the dangers of being turned: the accumulation of artillery against a special position can never be a very rapid operation. The subsequent "preparation" for attack is also lengthy. Hence there seems every reason why a moderate force, if able to secure a good defensive position, should hold its own for a much more considerable period than formerly. It will suffer more than formerly if it has to fall back after the attack has developed; but a body expecting reinforcement is not easily forced to fall back.

THE MARCH.

(*a.*) MODE OF FORMING THE COLUMNS OF MARCH WHEN A COLLISION WITH THE ENEMY MAY BE EXPECTED.

IT is necessary to deal in succession with various kinds of marches, all within range of possible collision with the enemy.

I take first the march during the opening of a campaign, before an enemy's position is accurately known, and when the two armies *may* be at a considerable distance from one another, though the distance is uncertain. In this case there seems no reason why armies should not, in ordinary country, be habitually preceded by large numbers of their cavalry, as was Napoleon's custom when he had sufficient to employ.* On more than one occasion he narrowly escaped being severely punished, because, either from not having enough light cavalry to spare for this work, or from having neglected the precaution,

---

* Operations of War, p. 423.

he was without the information with which they usually provided him.*

The brilliant use which the Prussians made of the same arrangement in the late campaign shows, as might be inferred from a consideration of the facts, that, under modern conditions, this advanced position has become more than ever the right one for all the light cavalry of an army.† The necessity for ample information as to the enemy, even if it be acquired at some risk, or even loss, has become more than ever pressing: the character of the action of cavalry on the battle-field has materially modified. For reasons

---

* Notably at Görlitz in 1813, and on the day before Montenotti. The most striking instance of the importance which he attached to the matter is the truce after Bautzen. That truce, one of the most direct causes of his fall, in so far as it was due to military calculations, seems to have been really wrung from him, more by the sense of the risks to which he had been exposed from being without cavalry, than even to his annoyance at the extent to which their loss deprived him of the fruits of victory. On the other hand, Captain Baring has pointed out that in the Ulm campaign the cavalry outposts were on the Danube at the very beginning of the great march.

† Captain Lahure speaks as if the Prussians had in this campaign invented the idea of "independent cavalry corps," an expression by which he appears to imply, amongst other uses, this of cavalry pushed forward far in front of the army. It is some consolation to learn from the other 'Conference Belge,' probably the most complete military monograph in existence, that the names of Murat, Kellerman, Seidlitz, are not yet in danger of being forgotten. One may be permitted to remark, however, that the difference in 1815 between the Prussian army and our own was, that our front was watched by an "independent cavalry corps," and that the Prussian was not; that we did obtain full information as to what occurred in front of us, and that the Prussians, to our mutual cost, did not, at all events, succeed in conveying early intelligence to us.

which will be considered under the proper head, it appears sufficient for purposes of battle to retain as reserve cavalry such a proportion of the arm as it will be necessary to throw as a deciding weight into cavalry contests, and to allow the whole of the remainder to be light horsemen. The wealth of every country in Europe has, moreover, increased, and, thanks to this and also to the development of the railway system, the resources available within given compass are much greater than was formerly the case. Experience has shown, that almost everywhere in Europe, if a wise system be adopted, and what is required is paid for, all that the country actually occupied by troops may contain is at their service. It is impossible to apply this on so large a scale as to make the supplies in the country sufficient for the wants of the main body of the army. But very large numbers of cavalry can be regularly supported by invaded territory, provided sufficient supplies for the following day or so are always carried, and the cavalry themselves are dispersed over a wide district. Moreover, the development of roads has become so great, that almost everywhere mounted men can, by passing along all available paths, and keeping up proper communication, search out nearly the whole of a country. Risk may occasionally be incurred by a few horsemen if they enter defiles unsearched by infantry; but as a rule in ordinary country, if the cavalry are in the habit of properly co-operating, and are pushed sufficiently to the front, there seems no reason why they should not, till they come into actual contact with the enemy, move in this manner. Most of the risk would apparently be obviated if a certain proportion of properly-trained mounted riflemen were sent forward with the cavalry on this duty.*

With the exception of necessary escorts, the cavalry required with each advance and rear guard, and a certain force for general duties, the whole of the light cavalry may perhaps be thus sent on. Their numbers can hardly be too large, as long as they are so employed. It is obvious to how great an extent the problem of arranging the order of a march is simplified by this. Mounted corps require great supplies of forage in addition to the stores necessary for infantry, and if these have to pass for a long time through the infantry columns, confusion and difficulty must result. There are serious reasons, however, why cavalry should not for a long time march directly in rear of infantry. Not only is one especial value of the arm, the rapidity with which it can bring back intelligence, lost, but the cavalry suffers seriously from having to follow at a foot's pace. It is scarcely necessary to refer to the familiar fact that, during the marches of the French

---

* Rossel, p. 146, quotes Nap. Mem. viii. : "Les dragons sont nécessaires pour appuyer la cavalerie légère à l'avant-garde à l'arrière-garde et sur les ailes d'une armée." The German General, p. 90, says, that " in the opinion of nearly all of those concerned in the fighting in the east of France, a far-reaching firearm was necessary for the cavalry." —See also Operations of War, p. 425.

in 1859, their cavalry horses were much injured on this account. Yet it would hardly be well, when moving within the area of possible collision, to allow the mass of the cavalry to remain on each day far in rear, and to come up at their own pace. It is to their pace in coming up that we trust when we leave them in rear of the infantry columns. To increase that distance, therefore, during the earlier stages, by the greater part of a march, would almost involve our being deprived of any assistance which cavalry might render during the phases of an unexpected action. All these difficulties are removed by the present arrangement. The cavalry march at their own hours, the infantry at theirs; and, in addition, not only are the trains relieved of much of the forage-supply for the cavalry, but also the latter will be able in many cases to arrange that large quantities of the more bulky kinds of forage are ready for such of the artillery as are near the head of the column. Hence, if this position be assigned to the cavalry, one somewhat advanced may without inconvenience be given to a considerable proportion of the artillery also.* The temporary defensive power acquired by relatively small bodies has been noticed. It appears to be a not unreasonable deduction, that artillery, protected

---

* Boguslaswki, p. 94. Nearly all the corps artillery "joins in action of *advance*-guard." Then artillery *en masse* moves up to effective range. So also German General, p. 245; compare also Borbstaedt, p. 345; Duke of Wurtemberg, p. 18; Prince Hohenlohe, p. 17.

by a fairly strong advanced-guard, runs little risk of being captured, provided the general situation of the hostile forces is discovered by the cavalry as soon as any of them arrive within thirty-six hours' march.

There are strong motives for assigning such a position in the order of march to the artillery. Any portion of an army which may be engaged more or less independently, and which, under orders given it, may require to seize upon and hold positions occupied by the enemy, needs for the purpose of attack a large proportion of artillery in hand. The enemy's skirmishers must be driven in before the artillery can come fairly into play. But in so far as the actual attempt to seize any position, large or small, is concerned, every fight, except under the rarest circumstances, necessarily nowadays commences by heavy artillery-fire. It is impossible for the infantry even to approach the lines of the enemy till the first fierceness of the fire of the defence has been taken off. The period, therefore, during which an artillery action is progressing, will be simply lost to the advance of the infantry, unless the artillery can be brought into action before it is necessary for the main body to approach. This preparatory action of the artillery has become so distinct a phase of the action, that it is essential to a considerable extent to adapt our order of march to it. Formerly batteries moved up with their division to the attack; and therefore, as the guns could always get into position when the line was formed, it was better that while a certain proportion of artillery was ready to protect formations, the remainder should come up when the general arrangements had been completed. But now, when it is essential to keep out of the *rayon* of destruction during all phases of a battle every portion of an army which is not absolutely required within it, it seems far better that only so much of the infantry as is necessary for the actual protection of the guns should be with them during the preliminary firing, and that the remainder should be ready to move up as the hostile fire somewhat slackens. They need only be ready to take advantage of the first impression made by the artillery. It must be remembered that now, when more than ever infantry requires the support of artillery, it will be somewhat less certain to receive a very efficient support from it at the moment preceding attack. It has, I think, rarely been the case in any actual battle of the past, that infantry would have been less efficiently supported if the guns had been of the present character, and had been somewhat further withdrawn. The reasons which determine the selection of battle-fields usually decide the question in favour of more efficient artillery action from a distance than from infantry range, even as to the mere fall of the ground. But there is for all that, no doubt, truth in Sir G. Wolseley's remark, that now more than formerly the extent of support which artillery will be able to render whilst infantry is assaulting, must depend on the character of the ground they are able to occupy, since they cannot move up so as to make use of canister. Hence, for this reason also, the length of time during which artillery can fire in preparing the way for infantry, without unduly delaying the general attack, becomes a matter of the greatest importance. Combining, then, these two facts of the greater security afforded to artillery by small bodies in consequence of the highly defensive character of the small-arm in operations of this kind, and the essential need of the presence of artillery very early in the action, the conclusion seems inevitable that a large part of the artillery ought to be placed near the head of the column. I should, in fact, propose to have in ordinary country, immediately in rear of the advanced-guard, and perhaps of a single battalion of infantry, the whole of the artillery of each division moving by a single road. Where a corps d'armée has to move by one road, it would seem best to have the divisional batteries of the leading division directly in rear of the advanced-guard.* Then, immediately in rear of the 1st division, the

---

* This was the general order of the Prussian advance on Gravelotte. And though there were details in which the artillery was certainly not well managed, the march appears to illustrate the form in which the principles, which both Boguslawski and the German General maintain that the war confirmed the Prussians, would be applied. Prince Hohenlohe in principle advocates much the same. It should be observed that the advanced-guard is here made somewhat stronger in infantry than would usually be the case, because the fact of the whole corps moving along a single road implies a country in which artillery would be exposed to greater danger, and be less immediately available.

reserve artillery of the corps followed by the artillery of the 2d division, then the 2d division and the remainder on the same principle. Any batteries of position artillery, however, and any guns kept as an army reserve, would follow in the rear of the whole column. Last but these would be the heavy reserve cavalry. Therefore, the order of march which would seem advisable, *if the whole of* each corps were by the circumstances of the country constrained to move on a single road, would be this:—

1. The light cavalry. Its extreme advance thrown out to a distance 48 hours ahead of the main body. Its main body, with horse-artillery, &c., half-way between the advanced-guard of the corps and its own advance.

2. Advanced-guard of all arms, including (order not here specified),

> One battery field-artillery.
> One brigade of infantry.
> Four squadrons.
> Some sappers—strength dependent on character of probable road.

First Division.

3. The other battery of field-artillery.*

---

* It will be observed that from the present constitution of our corps d'armée it is inevitable to violate, so far as the divisional artillery is concerned, the principle I have advocated of always, when possible, working with three or four batteries at a time. The Prussians have 90 guns

4. One battalion of Rifles.
5. One company of Royal Engineers.
6. 2d brigade.
7. One troop military police.

8. RESERVE ARTILLERY OF CORPS.

SECOND DIVISION.

9. A body which will, if necessary, form the advanced-guard of the division, should it ultimately be able to move by a separate road (told off and heading the divisional column, but simply forming part of it, and with no increased distance. The object of placing it here is to enable the division at any moment to move independently by a different road without interfering with the order of march, as would be inevitable if troops had in this case to be brought up from the rear to form it. Moreover, in case of the division coming into action, it is very desirable that it should have to support its own advanced-guard, not become intermingled with the general corps advanced-guard taken from the other division).

> One battalion of Rifles.
> One battery. Sappers, as with corps advance.
> One squadron. Whenever possible, however, this should be attached to the corps advance.

in their corps d'armée of 30,000 men — we have 78; the Prussians have two infantry divisions — we have three; hence we have two batteries per division to their four. The result is, that as one battery is required for the advanced-guard, the other must be left to move alone.

10. The other battery field-artillery.
11. One company of Royal Engineers.
12. 1st brigade.
13. 2d brigade.
14. One troop military police.

15. THIRD DIVISION, in same order as second.

16. The company of Royal Engineers attached to corps, and the equipment troop as ordered each day.
17. The pontoon-train, unless otherwise ordered.
18. The telegraph troop.
19. Any remaining cavalry of the divisions that is not sent far in advance following later, so as to arrive at the next halt at a convenient hour.
20. The heavy reserve cavalry, do. do.
21. Any position-artillery moving by this road.
22. Reserve ammunition.
23. Baggage-train, &c. &c.
24. One troop military police.
25. Rear-guard.

If collision with the enemy does not occur for some days after this order of march has been adopted, a different division would on each successive day take the lead, the advanced division of one day taking the rear of the column on the next. In this way alone does it seem possible to save those portions of the mounted corps, cavalry and artillery, which must form part of the column, from the injury caused them by having to be ready day after day at an excessively early hour. It is to be remembered that here those marches only are taken into account in which it is necessary to be ready for serious engagement with the enemy. In fact, such marches are not likely to be usually of very long duration. Every effort would be made to be prepared for the extra strain caused by them during the period when collision need not be anticipated; the lead of the actual column would then be habitually given to the infantry in order to save them, while, by the same means, the horses of cavalry and artillery could be fed at a convenient hour; and inspections being more satisfactory, there would be less fear of sore backs and of other mischief.

I have selected, as more conveniently illustrating the general principle, a march in which the whole corps has necessarily to pass through a district throughout which it is confined to one road, but may expect to meet the enemy in less enclosed country, as was the case with many of the Prussian corps at Gravelotte. But in almost all cases it will be better for each division to move by a separate road. A corps d'armée moving along a single road occupies, with the ammunition necessary for action, 18 miles. The longer a column is, moreover, the slower must be the pace at which a march is regulated, in order to avoid the increased delays due to inevitable losses of distance.

It may be safely inferred, therefore, that more troops will rapidly be available for action by moving divisions on separate roads than by crowding them one behind another, unless the roads are at least three miles apart. The increased temporary retaining power of an advanced-guard, if this be combined with ample and early intelligence brought in by cavalry, would prevent all danger of an enemy's breaking in between the columns. These, moreover, no longer require to be elaborately formed up before moving to support. They may be regarded as depots for feeding the advance. There was scarcely a movement of the campaign, in which the country at all permitted it, during which the Prussians did not move their divisions by separate roads. As long as the cavalry is 48 hours in advance, each division ought certainly to occupy a separate road, unless these are *very* widely dispersed. If there were roads within a reasonable distance of one another, it would be very desirable to move by brigades rather than by divisions, but our present supply-arrangements, being all divisional, hardly admit of this. When the army approaches within reach of collision, the cavalry will be obliged to contract their distance, but it will be essential for them never to lose sight of the enemy.* As a rule, however, the army, having advanced so far that its cavalry has met the cavalry skirmishers thrown out by the enemy, will be prepared either to assume some ad-

* Boguslawski, p. 89, points out that the cavalry were usually kept in presence of the enemy up to the last moment, and that this was invariably found advantageous. They were never at any time allowed to lose touch of him. He uses, p. 88, the expression that up to the very last they were required "Stets Fühlung zu halten."

vanced position in order to force the enemy to attack, or, if it be not in condition to attack the whole army, will seek to obtain some opportunity of assailing with advantage a detached portion of the enemy's forces ; or, lastly, the nature of the operation on which it has been engaged having drawn the enemy upon it, according to the intention of the movement, it will throw itself into as strong a position as possible and await attack. In the latter case the march is over, and the disposition need not here be further considered. But it is obvious that an army which desires to draw on attack must be ready to adopt rapidly on occasion one or other of the two first alternatives, to find occasions for them, and to act rapidly when these present themselves. In order to force the enemy to incur the losses of attack on a large scale, it is essential to be ready to incur these on the small scale by rapid local attack with numbers locally superior. Exceptional in its nature as was Alvensleben's action on the 16th of August at Rezonville, it suggests reflections of a much more general kind as to a possible advantage to be taken of the present condition of arms, more especially in order to arrive at the result for which, as I have maintained, an English army ought to strive. It is clear that the whole of, say, 120,000 men could not move with the same rapidity to strike a sharp decisive blow as a smaller body specially prepared to take advantage of any information that may be suddenly brought in by the cavalry. Assume, then, that the protecting cavalry have surrounded the army up to the latest possible moment, so that full information has been acquired as to the enemy's position and movements; also that an adequate advanced-guard has been pushed forward on every road, and all advanced-guards are in perfect correspondence with one another. The first essential would then appear to be—to have a force ready in hand for immediate action, powerfully made up of all arms, and sufficiently complete in itself to be able to take advantage of any favourable circumstances that may present themselves. Since a comparatively small body is with very great difficulty expelled from a position, the advantages of having ready for action such a force, powerful, yet capable of comparatively rapid movements, are obvious. There is very little fear that a position once so secured will not be maintained if the remainder of the army is within any reasonable distance of it. I am not advocating the old error of a mere concentration on an advanced-guard. That must result inevitably now, as formerly, where the concentrating army is really dispersed, in a triumph for the enemy in succession over every fragment as it arrives. But where an army is marching under the protection of a covering envelope of cavalry and of a proper system of advanced-guards, so that it receives full notice of what is going on before it, it enjoys the double advantage of being able to move on a very broad front, and therefore with great ease and rapidity, and of being able to concentrate to any extent that may be desirable before coming into collision with the enemy. If, therefore, an army of four corps be thus marching previous to any defined plan of action in detail having been arranged—the enemy's position not being accurately known—I should propose that one corps, prepared by every possible means for exceptionally rapid movement, be in advance, followed in a kind of second line of columns by the other three corps. In order, however, not to over-extend the advanced corps, or to contract too much the space taken up by the others, I would propose that the advanced corps, whilst furnishing advanced-guards for the whole immediate front, in order that all sources of information may be in the hands of the corps commander in the first instance, should not furnish them for the two flank corps on either side, which would find their own, and would, in fact, remain in a kind of echelon on either flank, extending laterally beyond the front occupied by the advanced corps.

There seems no good reason why corps should be always identical in their formation. I do not see what there is in modern conditions to alter the dictum of Napoleon, quoted by Rossel, p. 133,—" Il est bon que les corps d'armée ne soient pas égaux, qu'il y en ait de quatre divisions, de trois, de deux." It is quite true that during the late campaign the French seem to have suffered from this arrangement, as from so many others; but it would be absurd to conclude on that account that in every detail their scheme was defective and that of the Prussians perfect. I have assumed, in the above arrangements for the march, that the corps are all formed on the model of an English corps d'armée at present fixed.* But if differences be admitted in the constitution of corps, obviously it would be convenient to use that which, not necessarily numerically the strongest, had most divisions. If, as appears to be the present intention, a light division is formed, the corps to which it belonged, or the division alone, would be habitually in front. But this is clearly not *necessary* in order to enable such a form of movement as I urge to be carried out. All that is needed is, that there should be, so to say, a corps for duty. The great objection to the arrangement of having a special corps as a general advanced-guard is, obviously, that it destroys the corps' connection between each body and its advanced-*guard*. But I should propose to give the march a special protection as regards this point, to be noted under the next section. Moreover, what I now speak of is not, in the ordinary sense, an advanced-guard, but an advanced body, which may, without interfering with the general orderly advance of the army, be thrown forward into special action. It is impossible to lay down that either this or any

---

* Taken, as all other of these official details are, from the only work in which, so far as I am aware, they are published—Sir G. Wolseley's Handbook.

other form is advisable as an invariable rule. I only urge that it presents exceptional advantages under the present conditions of war.*

When the army is moving to attack an enemy in a known position, on the other hand, the precise order of march must be so arranged as to be appropriate to the special position of the enemy. In general, in this case of four corps, three would form the first line, and one would be held in reserve. The order of march of the several corps would not materially differ from that I have already given. When in actual presence of the enemy, as far as possible, if sufficient veritable dragoons—mounted riflemen—belonged to the cavalry division, these only would be left before the enemy. The remainder of the light cavalry would pass off, partly to the flanks, partly to the divisions to which they belong. Sufficient infantry from the advance must secure positions for the batteries.

* In a different form the same idea appears to have impressed itself on Captain Lahure, but he wishes to leave it to his "independent cavalry corps." Surely, however, cavalry is not the arm one would employ to take up defensive positions, unless, indeed, by "cavalry" mounted riflemen are meant. Now, no one has shown more forcibly than Captain Lahure himself how contrary a thing to all experience it is to believe that cavalry can retain their own best qualities, unless those who are to be mounted riflemen are as distinct as possible from those who are to act by charging. If sufficient cavalry are provided to form, with horse-artillery and mounted riflemen, a genuinely independent corps, in addition to the cavalry required for protecting the march of the army, that might be perhaps the best form of all of my "corps for duty;" but that cavalry and horse-artillery alone can now, less than ever, assume a defensive *rôle*, has been well established by Prince Hohenlohe Ingelfingen, if it needed proof.

Then the guns must be pushed on as rapidly as possible. Other details can only be fixed by special circumstances.

In very enclosed country the cavalry will not be able to lead. That duty must be intrusted to infantry. Though, from the extremely feeble nature of the enemy, it is perhaps impossible to depend absolutely on the experience of the Germans during the campaign of Le Mans and the West, it certainly would appear that powerful bodies of mounted riflemen may be safely trusted, and best trusted, to form the advance even in such country. Obviously, however, in these cases, where artillery can hardly be used effectively, the mass of it ought to be kept very much further back on the line of march. The Germans had to thank their enemy that they did not get this lesson pretty severely impressed on them.

Nevertheless, too much allowance must not be made for the character of the country. Where an enemy is known to be cowed or falling back, he is very apt to mistake properly-equipped dragoons for infantry.* The long Prussian march at the beginning of the war can hardly be regarded merely as pursuit, since the French army was by no means crushed. The success of that march seems to suggest that where the country is occasionally broken and occasionally open, the cavalry or dragoons should maintain the advance as long as possible, and the artillery

* Rossel, p. 148.

be as near to the front as safety permits. The march of August 1870 certainly did not lie always through open country.

The passage of long mountain-defiles has so often been a matter of importance that it requires separate consideration.

A few "mounted riflemen" should be sent well ahead down every possible road some miles in advance. Only so many, however, should thus go forward as may be necessary to examine the roads and keep communication with the rear. The advanced-guard would follow these. It would appear extremely desirable, from the experience of Trautenau and of Nachod, to provide an advanced-guard sufficient not merely to protect the march, but to clear the further side of the defile without delaying the general advance of the corps. Till the *débouché* from the defile has been cleared, the mere crowding up of the whole mass upon the front will induce confusion without increasing strength. The chance which is almost certain to be offered is this: the enemy will not meet in great force the advance of the column from the defile itself, because it will be impossible for him to know by what road the main force of the army will pass. The great object therefore is, not to allow the enemy's opposition to produce more confusion and loss of time than can be avoided.

Sufficient infantry must be in front to crown the

heights. A supporting body of infantry with a couple of guns would protect the immediate advance along the road. The remainder of the same battery, however, with only one waggon, might precede the next body of infantry, say one battalion strong, which would serve as support to the first advance. The reason of the distribution is this: the necessity for having guns as far in advance as safety permits has been proved by the experience of Trautenau; but more guns than two could certainly not be used effectively on the road itself; while, if it was necessary to support the infantry in clearing the heights, the guns would move out of the road much more easily, and come into action quite as effectively, if some space was allowed them in which to select their own way. It would be probably convenient, in such a passage, to have near to the front a considerable proportion of engineers, and they might safely follow these first supports. A small number of troopers should accompany the advance of the advanced-guard, in order to keep up communication constantly with the rear and with any dragoons sent out in front.

A considerable space should be left (say 400 yards) in rear of this double advanced body. Then a second support might follow, consisting of perhaps a battalion and another battery with three waggons. In order to have some cavalry ready to act on leaving the defile, the need of which is sure to be then felt, a squadron or two might, probably with advantage, lead the next body, which would form the main force of the advanced-guard, and be very strongly supplied with artillery, in order to force the passage on the further side. It would be absurd to fix accurately the force of this, since it will be largely affected by the nature of the defile, the character of the *débouché*, and the supposed force of the enemy on the further side; but in general terms it may be laid down that it should be sufficiently strong to force the *débouché*.

The remainder of the corps should follow at perhaps two or three miles' distance, a considerable proportion of artillery being near the front; because, in case the advance requires support, it will almost certainly chiefly need artillery, as was proved at Trautenau.* The distance from the advance of the main body is fixed by the consideration that it will not be safe to leave the advanced-guard without possibility of support during an action—that marching through a defile is necessarily very slow—that yet it certainly is here less safe than elsewhere to allow the main body to be attacked without sufficient warning — and that it is well, if possible, to leave time for clearing the defile. Obviously, in the case of a defile, the general principle that the advanced-guard should be suffi-

---

* By far the most perfectly arranged passage of a mountain-chain appears to have been Manteuffel's against Bourbaki.—Wartensleben, p. 11-50 especially. As yet, however, we have no details of it.

ciently far in advance to admit of the deployment of the column, does not apply.

All that can possibly be done to diminish the length of column till the defile is forced should be attended to. In order that the advanced-guard may be able to bring its force to bear as rapidly as possible, all waggons of artillery which can possibly be spared from their batteries should follow the rear of the whole advanced-guard. Without exception, all second reserve ammunition-waggons, all baggage, all ambulance, except the regimental ambulance, and all pontoon-trains, &c., should be left behind on the near side of the defile till the *débouché* on the further side has been seized. Whether the telegraph train should or should not follow earlier must entirely depend on the nature of the undertaking, and the degree of co-operation with other corps which is required.

PRECAUTIONS ON THE MARCH AND IN POSITION.

(*b.*) MODE OF COVERING AN ARMY ON THE MARCH OR IN POSITION IN ORDER TO CONCEAL ITS MOVEMENTS, AND TO OBTAIN INFORMATION OF THOSE OF THE ENEMY.

MANY of the precautions which will now in practice be of the greatest importance, have necessarily been dealt with in considering the order of march. In this part of the paper I intend chiefly to consider in what points of detail it is necessary to modify our present practice of outposts, and on what principles the detail should be arranged. The protecting body of cavalry thrown out in advance of the army will be the first, and during the early stages of the campaign the most important, of the means of precaution which an army can adopt. Obviously the cavalry must, when engaged on this duty, work to a great extent independently. Whether it will or will not be advisable to place the whole under one officer, or to leave the connection between cavalry brigades to be maintained by the several brigade commanders, must depend on circumstances. It is, however, exceedingly necessary in either case to take account in this arrangement, as in all others, of the connection between particular cavalry regiments and the corps to which they belong. I should propose, therefore, taking as an illustration the march of four corps, with one corps in advance of the others, that whilst the cavalry of the advance corps is thrown out along the front of the line of march, the duty be taken to right and left by the cavalry divisions of the right and left corps respectively. Thus each corps will have to its own front, and on the side on which it is most exposed, cavalry, whose reports will be furnished to their own corps commander as well as to the corps commander of the advance. The extreme outward patrols of the cavalry would be about 48 hours ahead of the main body, each cavalry division supporting its patrols at about a day's march in rear of them.

Each corps* would of course have its special advanced-guard. With each main advanced-guard of the flank corps a certain number of troopers from the extreme advanced corps would march, in order that all information might at once be sent to the commander of that corps. As, however, it would be necessary to have a general commander for the advanced-guard, and a sub-

---

* Except, in the case cited, the centre corps following in second line, which would obviously be protected by the corps in front, and need only keep communication with it and the flank corps.

ordinate commander of the outposts for each corps d'armée and division, it might be well that the reports of the cavalry belonging to the advanced corps should be delivered in the first instance to the commander of the advanced-guard of this advanced corps.

As I by no means urge this special order of march as always advisable, it will be sufficient to note that in whatsoever way the corps may be moving, the two most important precautions which it appears necessary to take for the sake of avoiding the risk of accidents, such as have so often occurred in the transmission of information, are these : first, to secure that the whole of the front shall be guarded by a body so far under a single officer that all reports may be in the first instance forwarded to him ;* secondly, to establish such a connection between the force following a particular road, and the advanced-guard on that road, as may avoid all risk of information not being immediately conveyed to the former. One difficulty among many which stand in the way of placing the whole advanced-guard of a large army under the charge of a single corps is, that if the several divisions occupy as

* I had not noticed when this was written that this is strongly Marshal Bugeaud's opinion (see Bugeaud, p. 72) as to the duty of each division, and that his editor (see Bugeaud, note, p. 183) had put forward the above. As it stood, it had been intended to rest on the notorious fact that the weak point of a division is where two brigades meet—the weak point of a corps where two divisions meet—the weak point of an army where two corps meet. It would seem advisable even to limit independently moving armies to such size as would restrict the space watched by their outposts to what one man could superintend.

many roads as possible, the corps in advance will be too widely dispersed for effective action. It is, therefore, that I do not propose that the actual advanced-guard duty of the whole army should be carried on by the corps in advance, but that from it patrols, both of infantry and cavalry, should be constantly passing to the advanced-guards thrown out by the flank corps of the second line whenever the army halts. Duplicates of whatever reports are sent back to the corps for which he does not furnish the advanced-guards should always be brought by his own troopers from other corps to the advanced corps commander. Any attempt to give the whole advance duties to the advanced corps would not allow the men of this corps to receive their fair share of rest, and would therefore also incapacitate them for those exceptional exertions for which they are likely to be called on.

Supposing that for special reasons it is more convenient to have all the corps simply marching side by side in a single line, it will still be necessary to appoint an officer to the general charge of the advanced-guards of the army, in order that any information that may be received by one corps may be known to some one who has the means of comparing it with reports from other directions, and of making sure that all that is necessary is at once transmitted to headquarters, and communicated to other corps. A properly-organised telegraphic service ought to enable the information which a commander-in-chief requires to reach him more rapidly by this means than by any other. All outposts and patrols should be under the command of the officer of the advanced-guard, whether of the division, corps, or army to which they belong. As far as possible the relationship between the several advanced-guards and the body of which they are the direct advance should also be maintained intact. But without some one responsible for the mutual co-operation throughout of the whole protecting body, it is difficult to see how risk of frequent misunderstandings can be avoided. Our tendency is to extend on the march over more ground than formerly. Such errors occurred even when extension was less.

Where the army is necessarily moving on a narrow front, the actual advance duties might be taken along the whole line by the advanced corps. If the corps are really organised unities, it will be inconvenient to break one up if it can be avoided, even should the army be very restricted as to the front on which it is compelled to advance. In case all advanced-guards are furnished by one corps, then there should be attached to each advanced-guard cavalry belonging to the corps whose march it directly protects.

I have dwelt at some length on these questions, because it is in the arrangement for larger combinations that our regulations are at present deficient. General Walker has pointed out that our English system of patrols and advanced-guards is, in the abstract, and so far as it goes, excellent; and that the less we imagine we can obtain the Prussian success by adopting their mere forms, the more likely we are to appreciate the spirit of what they have done. Minute details ought to be avoided. The fewer of them that are prescribed the better. What we require is effective practice on varied ground, and sound training. It is, however, important to notice, that at present no adequate arrangements have been made for the combination of the three arms on this duty. All that our regulations determine is the manner of forming the protecting bodies for each of the arms. An occasional note is no doubt added for each as to the manner of combining with other arms, but it is probably the natural result of the practical separation of the organisation of each from that of the other that the system is worked out not as a whole, but only as to one arm at a time, and to fractions of that arm.

Now there are, in fact, two distinct bodies engaged in an adequate system for protecting either a march or a position, unless the two armies are actually in presence of one another. That more distant from the army will consist of the larger portion of the cavalry, with probably nearly, if not quite, all of the horse-artillery also. It will have its own system of outposts and patrols in position, and its own system of advanced-guards and skirmishers in front of it on the march. For this body, so far as the detail on each road is concerned, the great necessity is, that as it possesses no defensive strength, and is peculiarly liable to surprise, only the smallest possible bodies should be exposed to the enemy's efforts. The advanced vedettes or skirmishers may be gradually supported by bodies still small, but capable of presenting a more serious resistance; these again by others yet larger, and so on until the main body of the cavalry is reached. The general form of this outer protecting body will in fact be as General Walker puts it—an open fan; or as Sir G. Wolseley has said of the nearer screen, an open hand. For the outer screen, however, the essential principle is that it should only be intended to afford a retarding resistance to the enemy. The extreme outer portion should yield at once. As this falls back, the whole will gather strength. It must then yield or not, according to the apparent strength of the enemy. As a rule, the smaller the detached bodies in front, the more easily will they both gather information and escape.

Our experience of the proper use of cavalry on this kind of duty is defective, for a very curious reason. Long as it has been the practice for armies to employ cavalry in this manner, there is no campaign as to which we possess records of the manner in which two armies, each thus prepared for the encounter, met one another. The American war perhaps supplies us with the nearest approach to an example; but their cavalry was organised on so different a footing from that of European armies, that it is difficult to draw conclusions from it. The question is, now that any satisfactory system is sure to be studied by every nation in Europe, how, when cavalry advance meets cavalry advance, is advantage to be secured? I have noticed already that Napoleon considered dragoons (mounted riflemen) to be the proper support for cavalry on outpost duty. They obviously supply the defensive element in which cavalry by themselves are deficient. A force the main body of which was well supplied with horse-artillery and with mounted riflemen, ought surely to be more ready for all emergencies than any other that could be thrown forward into this advanced position.

The Prussian experience appears to have forced on them a strong conviction of the necessity of a powerful proportion of mounted riflemen possessed of a really effective infantry weapon.* If this is so, it would seem to be a legitimate deduction that these men should be, as Napoleon found advisable, equipped as completely like infantry as possible. Further, that instead of moving by cavalry fours and sections, they should when on foot adopt a

* This seems the fair inference from the assertion already quoted from the German General, that all engaged in the campaigns in the west became convinced of the necessity of a far-reaching arm for those cavalry who had to fight on foot. The superiority on foot of infantry that has been mounted to dismounted cavalry has been too often proved to need fresh discussion. See, however, 'Modern Cavalry,' p. 7, 20, 124, 132, 139. On the other hand, the superiority of a cavalry trained to believe in the effect of their charge over cavalry quite otherwise taught, has been again and again illustrated: see especially note to p. 155, *infra*.

regular infantry drill. The action of bodies supporting cavalry has so frequently to be moral rather than forcible, to impose on the enemy rather than to injure him, that every advantage that can be taken of the chance of leading him to believe that the more decisive portion of the army is arriving should be employed. At the same time the principle must be laid down, that, except under the rarest circumstances, these cavalry advanced bodies are not intended to fight, but to watch the enemy. It is only when touch has been lost that it must be recovered at all cost. For this reason, therefore, the horse-artillery and the greater portion of the mounted riflemen ought to be kept with the main body of the cavalry advance, not pushed further forward. The exact distance apart of the several fractions of this extreme advanced body, or the number of steps by which the force should become increasingly larger as the main body is approached, can hardly be laid down. This depends on an infinite variety of circumstances. As a rule, however, it may be said that from the extreme advanced patrol to the main body of the independent cavalry a day's march should intervene.

The inner protecting body is designed for a somewhat different purpose. It is distinctly intended to afford not only information, but direct protection, to the main army. It is formed of all arms. It is with reference to this force that the expressions so frequently employed as to the use of the several arms on "outpost duty," must be understood to be intended to apply. Here the cavalry will still be the best force for the advance in the daytime and on open ground. The infantry will best perform the same duty at night and in enclosed country. The artillery are only placed on outpost duty with this body at all when the ground admits of their effective action. When they are so sent forward, it will not be to the positions which they are intended to occupy in the event of attack, but to the main body of the outposts of that part of the position. When time permits, epaulments will be prepared, and distances calculated at the various points which artillery is intended, if necessary, to occupy. By this means a smaller number of guns at the outposts can be made available for the same extent of ground.

It is essential that this outlying force should be entirely distinct from the main body. It would usually be formed from the advanced-guard. In position, supposing some such scheme as that suggested by Sir G. Wolseley, for the detail to be carried out, the most important point to be decided would seem to be that of the distance from the main body to be guarded and watched. The danger which is now incurred by an army whose outposts are not pushed sufficiently far, was illustrated again and again during the war. The panic produced by guns opening suddenly, as they now can, from immense distances, where the enemy's position has been reconnoitred beforehand, is sure to be most disastrous.

The best remedy would appear to be to have small detached posts well pushed forward on the possible lines of an enemy's advance which threaten most danger. The system long since proposed in this respect by Marshal Bugeaud seems even more adapted to our present condition of warfare than to that of his time.

It is to be specially noted that night-attacks were quite a prominent feature of the late war. Again and again positions, which the French had won at great cost during the daytime, were recovered by the Prussians with little difficulty in consequence of the careless manner in which the night-duty was carried on.

If the country around has been thoroughly searched to sufficient distance by cavalry during the day, there ought to be little difficulty in selecting positions for concealed detached posts of observation which an enemy must pass in order to approach *rapidly* and *quietly*.

The next point of importance appears to be to apply more thoroughly our system of patrolling rather than of mere sentries in camps. The principle is as distinctly laid down in our cavalry regulations as it ever was in the Prussian, that patrols are far more essential than sentries. A system of patrols has been also adapted to our infantry outpost system.

All that is required in this respect is, that it should be thoroughly understood that for the livelier, the more active phase of war, on which we have entered, the essential thing is to be possessed of the fullest possible information. We must not await quietly in our camp, under certain protecting sentries, the moment of the enemy's advance. Lost time is now too costly for that.

It seems unnecessary to go into the exact form for carrying out the process of arranging the radiating fingers of the "open hand," since no form can be prescribed for any further purpose than that of illustrating how such an arrangement *may* be carried out; and this has been already done by Sir G. Wolseley (p. 181, 182). It should, however, be noticed, that it is not very clearly determined in our regulations whether "the reserve" and the inlying pickets are or are not the same body. It can hardly be too strongly insisted on that the duty of the inlying pickets is merely that of keeping up the necessary vigilance over the sleeping camp. The outposts ought to be an entirely independent body, communicating with the inlying pickets when necessary, but entirely distinct from them.

There seems to be no object here in going into the question of the ordinary duties either of an advanced-guard commander or of a rear-guard. The principles in regard to them are well established, have undergone no essential change, and can scarcely be discussed briefly without leaving a false impression. It may, however, be noted, that from the increased difficulty of regaining lost ground, an advanced-guard ought to be more chary than formerly of yielding to any but a definitely superior force. A rear-guard, on the other hand, whilst taking full advantage of the increased retaining power of a small body over a large one, needs to be more than ever careful to abandon its ground in good time, lest its retreat should be cut off, or it should suffer terrible losses in falling back.

Some principles of the Prussian army for forming advanced-guards on the march appear worthy of adoption. We have no need to imitate their forms. Their first support to the flanking parties and advance is of much the same organisation as our support,* and is called a "vortrupp;" but immediately in rear of this they have a second support rather nearer to the first support than to the main body of the advanced-guard, and called a "hauptrupp." This is a body somewhat stronger than the vortrupp, especially in infantry, and with a fresh number of guns,—whilst, though the vortrupp has fewer infantry, it has some cavalry attached to it. The idea is obvious. It is intended that as soon as the vortrupp is engaged in action, a powerful supporting body should be at hand to enable the advanced portion of the guard to sustain the fight till the main body has come up. This is clearly adapted to those conditions of modern war-

* Sir G. Wolseley, p. 216.

fare which make it so difficult to regain lost ground, so comparatively easy to hold for a short time against superior numbers a position once secured. It is further consistent with those circumstances which render it far more dangerous than formerly for an army to be caught on the march even by small hostile detachments. Our advanced-guard is obviously intended not to fight. This principle, undoubtedly applicable in past times, may often be sound still. But there are times when it would be very dangerous, and the opportunity of acting effectively ought to be in the hands of the commander of the advanced-guard.

The principle that the distance of the advanced-guard from the main body should be such as to admit of the latter's coming *into effective action* is as sound as ever. But "deployment" is a word of somewhat doubtful signification now, since it may be questioned whether battalions will ever again be deployed, in the old sense of the word, in action. The distance so assigned is still, for want of a better principle, convenient as a guide. But it must be remembered that it is now more than ever necessary to be ready on occasion to sacrifice the less for the greater, and the sole limit to the degree of advance advisable is the danger lest advanced troops should be beyond the range of effective support, and so expose the head of an unguarded column. As a general rule, the Prussian principle in forming an advanced-guard seems unmistakably the best, that each part should be so far from the next following, that the latter is not exposed to *effective* infantry-fire. The latter they now fix at about 400 yards.

There is an immense amount of detailed precaution to be applied to all outpost duties into which it seems unadvisable to enter here. To select special points to be attended to, is to throw into the shade the importance of others. To give all with all the variety of circumstance which presents itself would be here impossible.*

---

* Details are given both in Sir G. Wolesley's book and in Sir C. Staveley's. De Brack and Bugeaud are full of them.

THE ATTACK.

(*c.*) MODE OF FORMING, COMBINING, AND EMPLOYING THE DIFFERENT ARMS FOR ATTACKING AN ENEMY IN POSITION.

THE great difficulty in making deductions from the late war as to the details of a proper system of attack is due to a peculiar cause. It is scarcely possible to show what this is more clearly than by comparing the broad statement of Major Tellenbach, that in all fighting the greater proportion of fire is unaimed, with the observations of the Duke of Wurtemberg as to the fact that the French hardly made any effort to secure aimed firing. The question is, how far can the data on which the Prussians worked during the late campaign be counted upon as permanent, and what will be the differences which will present themselves when troops have to be faced who have as carefully worked out the problem of the right use of the breech-loader in defence as the Prussians have themselves done? From this point of view very much the most interesting evidence that we could receive would be a thoroughly-thought-out statement as to the solution of the tactical question from the French side. Unfortunately it is impossible, with such a close accordance on the main points as exists between the various writers who have lately discussed the matter in Germany, not to feel that we have as yet received nothing from the French which can fairly be taken as a real study for the tactics of future attack. One thing, however, may be safely assumed. In future we shall not have, as the Germans had, to attack positions whose defenders waste their fire at useless ranges. The principle which Count Moltke pointed out as the right one on this subject eight years ago, will be doubtless largely adopted; picked shots will be selected to fire at very considerable ranges, in order to disturb the regularity of formations, and to make the action of artillery dangerous. Artillery will be aimed, with incomparably greater accuracy than the French displayed, at any bodies that may present satisfactory marks. But the intense severity of fire will be reserved for ranges at which the power of the rifle can be made to tell effectually. Thus the fire at 1000 yards or more will be much less severe than the Prussians found it; while at ranges of from somewhat below 500 yards to the position itself, the fire will be far more deadly even than that to which they were exposed. Now, the exact formation which troops will in the first instance assume, must necessarily very largely depend on the length of distance over which those formations must be adopted which are now essential for the actual advance against a position. However, if it becomes evident that an assailant has presumed on the defenders' reservation of fire too much, and has presented himself within ranges at which fire, though ineffective against loose formations, is effective enough against close ones, we may probably accept the Prussian experience as at all events the best we are likely to get, even as to the ranges at which modern fire renders all close formations for rigid movement impossible. That distance, in so far as large columns was concerned, seems to have been about 2000 yards. It may be assumed, then, that the march to battle will now terminate at some distance not much nearer than this to the enemy's position, and that the manœuvres of fighting must be considered from that distance at least onward.

Here, however, we are at once met with a most marked contrast to the past manner of beginning a battle. Formerly the march to battle preceded a careful arrangement in lines of all the troops that were intended to take part in the combat, or of at least the major portion of them. Napoleon's famous parade before Waterloo was quite in keeping with the then current phase of war. As a rule, the general who had begun an action before his troops were up, would have exposed himself to such disaster as that of Dennewitz.*

* Though no doubt that defeat was largely due also to want of sufficient precautions on the march.

But the Prussians practise and maintain as a principle now a quite different mode of action. The moment an enemy's skirmishers have been driven to a distance sufficient to enable their artillery to approach, they bring forward heavy masses of guns, which they employ at first at long ranges, to take off the intensity of the hostile fire. Then, under protection of these, their own advanced-guards push in the enemy's skirmishers, and secure yet more advanced positions, to which the artillery move rapidly forward, and bring effective fire to bear on the main body of the enemy. At Gravelotte they undoubtedly carried this too far, and lost guns in consequence. But it must be remembered that they had to discover for us experimentally, no one having then done it, what was the distance from infantry at which it was possible for guns to be worked.* Moreover, though in this instance the attempt was distinctly a mistake, it must be admitted further, that the Prussians have adopted a principle very new to military rules. They have decided that, in order to win victory, it is well worth while to run the risk of losing guns. The point, no doubt, might sometimes be pushed too far, and to a subordinate commander the exact nature of the stake for which he plays ought to be very clearly apparent when he ventures on the risk. But whether, as a matter of principle, it is better to govern Europe than to

---

* See the most careful working out of this question in Lieutenant Hime's Essay.

preserve a point of military punctilio, is not a matter open to question. It has been already necessary to discuss at an earlier stage the difficulties involved in this artillery operation, and what appears to be the right method of meeting them.

The cause of the advance towards the second distance, which is put at about 1600 paces or rather less, needs to be noted. The difference in the effect produced by artillery firing at 1600 paces or at 2500 is not so important as this. In order to be able to act as far as possible in an oblique direction along the main line, artillery has necessarily to fire against points more distant than the nearest part of the enemy's position, since this is directly opposite to it. This must increase the range considerably; and hence, since at very long ranges artillery-fire cannot be highly effective, because of the difficulty of seeing the objects aimed at, the motive for making one long second advance into favourable positions is clear enough. The chief precaution recommended in favour of the artillery appears to be the rapid throwing up of epaulments,* and the protection of the flanks of the batteries by troops specially assigned to that duty. One fact, however, is forcibly brought home by a consideration of this question of the independence of artillery. The more necessary it is to be ready on

* This question of the rapid throwing up of epaulments has been carefully worked out by Captain Macdonell, R.A., Artillery Institution Papers. Whether, however, it would usually be advisable for the gunners themselves to do the work may be doubted.

occasion boldly to push guns forward, the more essential is it that they should be protected as far as possible by troops whose only care it shall be to look after them. Of all classes of escorts for guns, the mounted rifleman* appears to be in every way the best. Since it is so essential, as the experience of war has proved to the Prussians that it is, that guns should manœuvre independently of other arms, can that condition be fulfilled satisfactorily otherwise than by making a proper dragoon escort form part of every battery? The most unpleasant of all duties would be taken away from other troops. The necessity for a general's constant fear lest his guns should be unprotected would be removed. The confusion which must ensue if artillery which no longer accompanies infantry is protected by an escort from some regiment, the headquarters of which are taken by the circumstances of fight before the end of the battle miles away from the guns, would be avoided. I scarcely see how, when artillery is to be pushed forward at the beginning of a fight into very advanced positions, escorts can be dispensed with; though it is well that, as a consequence of the greatly-increased difference between the ranges at which infantry and artillery now act, the question whether escorts are of much value should be raised. No escort with a battery will prevent guns from being disabled, if the enemy's infantry gets within effective range of them. On the

* See a discussion of this subject in 'Practical Artillery.'

other hand, an escort holding ground whilst artillery falls back will be more powerful than formerly, especially against cavalry, who are likely frequently to make bold efforts against guns when these are far separated from infantry. This seems a kind of work which could not be carried out without constant practice of escort and guns together.

Such being now the necessary prelude of battle, the infantry meantime moves up in columns of manœuvre* till advance in solid formations ceases to be possible. With us line of quarter-columns would probably be, under ordinary circumstances of ground, the most convenient formation. The problem is, in what way can the infantry advance to assault the position from the point where quarter-columns can no longer move?

The advance over the first part of the remaining ground where the enemy's infantry-fire was not very severe, was effected by the Prussians in company columns. An advance of companies in columns of sections would, without unnecessarily altering our drill, appear to meet the purpose perfectly. The Prussians found themselves able to maintain their advance in this order up to about 1000 or 1400 yards. They urge, however, that neither in this case nor

* The Prussians usually found columns either at deploying intervals or at thirty paces most convenient. No change of importance, therefore, was made in this respect.—Boguslawski, p. 69 ; Translation, p. 75. This, however, is expressly attributed to the defective character of the French artillery-fire.

in any other is it possible to lay down rules as to the exact distance at which one kind of formation becomes impossible and another must be adopted. At 1000 paces or thereabouts they found themselves compelled to extend one section (the "zug" being, however, a third, not a fourth, of a company) in skirmishing order. From this moment the constant tendency for more and more men to become involved in the skirmishing lines began to make itself felt. Very often, almost immediately after an action had begun, the whole of a regiment (3000 men) had without orders extended itself into simple lines of skirmishers. The question, therefore, is, how to introduce order into the movement. The Prussians meet it boldly. They assume that it is impossible any longer to keep the whole of the supports behind the skirmishers in any formation rigidly preserved. The only method is to train every man and every officer to understand that victory can only be won by co-operation, and by loosening the ties of discipline as little as possible. That done, each petty unit must take its orders as to the degree of looseness to be adopted from the man in immediate charge. He must receive his general indications from those above him. Each must endeavour to place himself effectually under orders again as soon as possible. Certain other principles appear to have been clearly brought out. In consequence of the extreme deadliness of front attack, every effort must be made at first to edge round towards one or both flanks. This will, in fact, be done instinctively, whether we make effort for it or not. It has been established clearly that that has occurred on the field of battle which had been previously noticed at practice—viz., that the tendency of men in firing is to aim at the centre of an object; hence that the flanks are the least exposed portions of each unit, and that, as a consequence, all supports ought as far as possible to be brought up behind the flank. These supports should, in fact, be echeloned from the body which they are to aid. The chief result of the arrival of more men is to push forward the skirmisher line,—the immense effect of even a few more men thus brought in being one of the means of action constantly to be taken advantage of. It is said that supports may safely be kept further back, since the line of skirmishers is always strong enough to protect itself. The temptation to use up supports too near at hand is very great.

So loose an order of fighting, however, with all precautions, necessarily leads to the intermingling of whatever troops are brought into action; and as in sending a general army reserve into the fight corps must often be thus supporting one another, it is found that even different corps get intermingled. Nevertheless the extreme importance of preserving tactical unity and cohesion is insisted on all the more strongly on this account, and every effort is made to retain it.*

* I prefer to leave this in the text in this somewhat vague form.

In endeavouring, then, to apply these principles with as little alteration as possible to our drill, I should propose the accompanying form (see diagram 1) for the general idea of the attack of a brigade on a position seriously shaken by artillery-fire. I assume four companies to the battalion, and four sections in a company, according to the organisation I have proposed already.

Two battalions of the brigade will be in advance, the remaining battalion will be held *as a brigade reserve*.

Of each of the two advanced battalions one company will be in advance: two companies will be in support in echelon, one behind either flank of the advanced company; one company will be kept in hand by each battalion commander as a battalion reserve.

Of the advanced company of each advanced battalion one section will be thrown into skirmishing order at 1000 yards from the position, or wherever it is found to be necessary. The other two sections will follow at first at about 100 yards in rear of the lead-

The writer, who has most carefully discussed the questions which have presented themselves under his own eyes, and the Prussian authorities, are apparently directly at issue as to the best method by which these results may be secured, so far as training is concerned. The Prussian regulation idea is, that in order to preserve tactical integrity, each section should close in so as to leave room for the supporting one to come on its flank. Boguslawski maintains that this practically is not possible, and seems to wish that we should purposely introduce disorder in order to be prepared for it. Surely Colonel Gawler's is the better solution—to give men ample practice, to let things take their chance at practice, and to accustom men to be ready for whatever happens.

ing section, and will move in such formation as the ground permits. The distinction between the mode of advance of these supports and of the skirmishing line will be this: the duty of the skirmishers will be to bring fire constantly to bear upon the position, and to move steadily onwards: the duty of each of the supporting sections will be to seek cover sufficient to enable them to remain, as far as possible, under the immediate orders of their officer. Each section will be pushed on and into the fighting line only when the first force of the line in front appears to have been expended, and more support is required. It would be impossible to lay down rules for distance in any degree absolutely. But the above explanation and the accompanying sketch are intended to illustrate the manner in which the several principles I have spoken of are applied. The whole movement is based on the principle of small units adapting themselves to the ground. The first supports should be as near as they can safely be placed to the first line. The strength of the line in front is the justification for placing the battalion and brigade supports far back. The second brigade would be kept in close formation well in rear. The object throughout is to present to the enemy's fire (to reverse the Duke of Wurtemberg's complaint against the French) a deep order of battle, but by no means dense formations (p. 32). A somewhat converging attack is provided for by the separate advance of the two battalions. The connection between them is secured, as we are assured that the efficacy of modern fire allows it to be, by troops echeloned in the intermediate spaces. All supports are brought up behind the flanks. The enemy's attention is divided as far as possible by troops appearing "from different directions at different distances and in different formations," while, at the same time, every effort is made to preserve the tactical integrity of units as long as possible, by placing supports from each company behind the advance of that company, from each battalion behind its advance; finally, by arranging to have each brigade supported by a brigade of its own division. The mechanism of the attack is intended to be that shown by the Duke of Wurtemberg to have been so successful, "the rapid change from open to close order directly the most trifling cover" admits of the rallying of a section, subdivision, or company. On the other hand, every advance over open ground, whether of supports or of reserves, will, when necessary, take place in widely-extended skirmishing lines moving on "like ants."

The detail supplied by our field-exercise book, though introduced before skirmishing had attained its present importance, appears to be compiled on as sound principles as if all the experience of the war had been before the writers. Certain obvious modifications of expression will be required in the part devoted to battalion drill, *now that skirmishers are not* *so much designed to protect columns, as columns are designed to supply skirmishers with successive supports and reserves.* It may be questioned whether it is wise to have supports "in rear of the centre of their own skirmishers," after the evidence on that subject supplied by Major Tellenbach.* Moreover, there is another point alluded to in the drill-book, the facts as to which require some consideration. It is there said that the formation for supports, when opposed to artillery-fire, ought always to be "in line." Now undoubtedly this is true so far as the experience of former wars is concerned. It is due to the fact that it is the easiest thing in the world to lay a gun correctly as to direction, but extremely difficult either to lay it correctly as to range, or to insure in this respect accuracy of fire. Hence there is very little use in opposing a comparatively slight front to artillery-fire, and slight depth is of more importance. But there are few defects of any arm which may not be met if the means of meeting them are properly thought out. Now the Prussians meet this difficulty of artillery in three ways. First, By habitually concentrating so

* Though the same phenomenon had been previously observed at practice. Attention was drawn to this before the war by Captain C. B. Brackenbury in 'Foreign Armies and Home Reserves,' p. 212, the article having originally appeared in the 'Times.' The passage, very elaborately worked out so far as the attack of cavalry upon infantry is concerned, is very interesting, now that so much of tactical action is at present based on the fact referred to. It may be questioned whether, now that tactics will be somewhat adapted to take advantage of this weakness of fire, practice should not be employed to check it.

many guns that if a certain proportion obtain the correct range, that is enough to secure the destruction of the line at which they fire. Secondly, By telling off certain pieces to fire at a little less than the supposed range, others at a little more. Thirdly, By devoting their whole training rather to the question of rapidly estimating range than to mere correctness of line. If to these means be added proper application of modern range-finders, it may well be questioned whether it will be possible to maintain any line formation even when artillery is the arm opposed.* On the whole, therefore, it seems impossible to say more on the subject of the formations to be assumed, than that they must be based upon a thorough knowledge on the officer's part of the manner of action of

---

* See on this subject Stoffel, p. 339 to 343. This paper has recently been translated into the Artillery Institution papers by Lieut. Hime. On the effects of artillery on line see also the staff officer's letter quoted on p. 29, *ante*. That may be perhaps regarded as exceptional, but it was not so considered by the French officers. They appear to have looked upon it only as a fair example. There is another point of the greatest importance to be gleaned from the same incident—viz., that whilst the infantry engaged in any particular action is now so more than ever absorbed in it that it must be left to work out its fight where it is, artillery, on the other hand, has by reason of its increased independence become more than ever available for suddenly changing the object against which it is to direct its fire. This is, in fact, the justification for employing artillery to so great an extent during the early stages of an action. Artillery is only diminished in efficiency by being engaged in proportion as it is either actually destroyed or as it becomes difficult to supply it with fresh ammunition. The latter point, no doubt, will require constantly to be attended to if artillery is to be so long firing. But as to the former, the more guns are employed in keeping down the enemy's fire, the less will be the destruction to the artillery.

the arm opposed to him, and applied to the special ground and circumstances before him. But the two essential points—the necessity of making every effort to obtain complete command over the fire of the men, and the leaving of the utmost latitude as to details in the hands of the officer on the spot—are applied in the drill-book at every point. It is not in the attempt to secure, but in the means for securing, effective control over the movements and fire of the men, that we fail. Unless all the evidence we now possess is worthless—and it is singularly in accord on all these points—it will be as utterly impossible for officers each to fulfil the work which our present organisation entails on them in skirmishing in modern fight, as to manœuvre a crowd collected round St Paul's.

It is careful previous adaptation of all throughout, by organisation and training devoted to the one end of adaptability to modern fight, that can alone supply the deficiency. During the late manœuvres skirmishing lines fell back, not behind other successive supporting lines, but yielding step by step to the opposing skirmishers, though the manœuvre is essentially contrary to the principles laid down in the drill-book, and will certainly result always in the whole line retiring when that man retires who most likes and wishes to do so. But it is not possible it should be otherwise as long as to the men themselves are assigned the contradictory duties of looking almost solely to the enemy, and yet of attending to the general course of the whole movement. The more small knots are placed under the authority of non-commissioned officers, and the whole system is built up from that point onwards by a regularly progressive extension of the command, the more complete will the effective power of manœuvring become. We are assured that all orders actually under fire can now be given only by having them passed on from group to group. The duty of the non-commissioned officers will not be so much to enforce authority, as to be constantly on the look-out for general directions, and to supply them to the men. It will be observed that in the assigned organisation (p. 60, *ante*) 4 sergeants' squads are made to form a section; this was done with the intention of frequently employing each section in the same kind of formation as is here shown for the company or battalion—viz., one sergeant's squad in advance, one supporting either flank, one in reserve. It is not shown thus in the plate, because the intention is that the lines should simply represent sections *in any formation* which the ground suggests. Since, finally, no success can be achieved that will be permanent, except by infantry placing themselves in the positions that have been occupied by the enemy, some means of gathering for a final advance must be arranged. It is difficult to understand how any such movement can be effectually carried out without an organisation so perfect as to maintain order amid all the tendency to complete disorder.

Supposing 8 companies must be reckoned to the battalion, the only hope of keeping them in any harmonious working would seem to be to place each wing severally under the command of a field-officer, and either to keep one in support of the other, or to make with the battalion instead of with the brigade a converging attack. But the use of wings was not very successful during the late war.

A variety of questions are still in dispute, and are likely to remain so, the fact being that the answer to them mainly depends on the circumstance of the moment and the man who has to solve them.

Boguslawski considers no longer possible the "ideal" *rôle* of cavalry on the battle-field, that of breaking in on the flank of an army or a large corps and rolling it up. The German General, on the contrary, judging from the example of the Rezonville charge, thinks that a body of cavalry supported by infantry may still break through troops and destroy them even if the body they attack has not suffered much beforehand. Surely it cannot be assumed that it has ceased to be possible to employ cavalry in such a manner as they were employed at Rezonville.* As the General

---

* Unless, indeed, you attempt to make the same cavalry engage in this grand *rôle* and in foot fighting, then the habits to which you train your men are essentially contradictory, and your cavalry will certainly "fight indifferently" on foot or horseback. Compare Borbstaedt's splendid description of that wonderful charge with the positive evidence supplied by one after another of the American generals to Colonel Denison as to the extent to which their cavalry, trained to believe in the effect of fire-

says, if they succeeded, as they undoubtedly then did, in obtaining a quarter of an hour's respite for the infantry, they probably thereby exercised a decisive influence on this part of the war. For he admits what has hardly been acknowledged on the German side before, that the 3d corps was completely exhausted, and would have been incapable of further exertions had not this relief been given them. It must, however, be remembered, that the cavalry so employed will as an effective force be almost destroyed. The cost will sometimes be needful, but it ought always to be counted.

On the other hand, the uses of cavalry on the field of battle continue to be manifold. Shaken infantry, infantry disorganised by the passage of an obstacle, caught by cavalry issuing unexpectedly upon them—infantry, when falling back from unsuccessful local attack—will almost certainly be crushed by good cavalry. A mere attack, however, on the flank of a body engaged with other infantry, if it has not lost many of its terrors for the assailed infantry, has become very dangerous for the cavalry. For the depth in proportion to numbers of infantry formations having so greatly increased, the cavalry are sooner or later almost certain to be exposed to a ruinous fire from the supports wherever they may strike the skirmishers. Nevertheless, the less cavalry are supposed

arms when dismounted, were cowed by firearms when they met them in the hands of other cavalry.

to be able to act, the more numerous will be the chances presented to them. Amongst other facts, the tendency for the general situation of the artillery to become very widely separated from that of the infantry must afford cavalry opportunities for striking at the guns either by wide outflanking movements or by seizing local chances. Hence a certain proportion of light cavalry ought to be present with each division, and large bodies will be required for general safety.

On the whole, however, it appears clear that in the attack of a position, the chief duty of the assailant's cavalry will be to protect the army from any widely-extended cavalry movements attempted by the defenders, in order to threaten flanks and rear. Hence, since in these encounters of cavalry against cavalry, the best cavalry will undoubtedly prevail, and it seems exceedingly unlikely that now more than before mounted riflemen can be made into good cavalry, it would seem as necessary now as it was in the time of Napoleon, that while light cavalry on outpost duty are supported by mounted riflemen, the light cavalry on the field of battle should be supported by a powerful body of cuirassier regiments. The more necessary it is that the light cavalry should be freely employed during the larger movements, the more essential it becomes to preserve intact for the field of battle the force of heavy cavalry which ought to rule the day in the cavalry action itself.

As the battle draws to an end the chances of the cavalry will now as always become more numerous. A powerful force must at all times be in hand to pursue vigorously in conjunction with large bodies of artillery.

The alterations which have occurred in the relationship of cavalry and horse-artillery, and in their mutual action, are not of very great importance. They have been well brought out by more than one writer,* and may perhaps be summed up thus. Horse-artillery should not close as a rule to case distance, as it would formerly have done; but having advanced rapidly ahead of the cavalry to its first position, it should maintain that as long as possible. The cavalry should endeavour not to cover the battery's front. If the guns move till the action is decided, it should only be to get into a situation on the enemy's flank. The most difficult task for the artillery commander will be to judge how to move when the cavalry action is decided either in favour of or against his own cavalry. In all cases the safety of the guns should depend not on special escorts, but on the mutual co-operation of the two sister arms. For the artillery, the essential is that movements should be as few and as rapid as possible. The Prussian experience, moreover, seems more and more to tend to restrict the use of horse-artillery to

---

* See Captain Kitchen's paper in the Artillery Institution papers, and the Prince of Hohenlohe Ingelfingen's elaborate working out of the whole question as to the two arms.

association with cavalry, and to employ for all other purposes—reserve included—a field-artillery as mobile as the heaviest available weight of projectile will allow, and as heavy as the greatest possible mobility will permit.

A very important question next to be considered is decided in two directly opposite senses by the two writers who have recently discussed it. Is it or is it not necessary, as a rule, to engage the enemy along the whole line whilst concentrating effort against a particular point? The German General says expressly that (p. 246) fighting all along the line like that of Gravelotte is not usually a matter of necessity, as it was not in that case. Boguslawski says, on the other hand, that the front engagement is a necessity (p. 56), in order to hold an enemy whilst the decisive attack is being made. There is one cause which has, so far as I know, not been pointed out, why an attack will require now to be supported by troops on either side of it to a greater extent now than formerly. Formerly, when from the time that troops had fairly come within the fire of the defenders to the actual attack, the period was not long, the really important part of the fighting usually took place much nearer to the enemy's line than it does now. As, then, the heads of the columns advanced to the attack and approached rapidly the assailed position, the fire of the nearest points of the line became more and more hampered by the awkwardness of the angle at which they had to fire. Now, however, that instead of charging directly over the ground, the skirmishers have gradually to acquire a superiority of fire over the defender, the greater part of the fighting will take place at a distance from the defender's position, such that the men on either side of the part actually assailed will be able to bring a much more effectual fire to bear. At the same time, the increased range of the weapons will enable a much larger number of men on each flank of the position to join in pouring fire on the assailing skirmishers. As every additional weapon brought effectively to bear is so much more important an element in the question than it was formerly, this cause will necessarily tell very severely in bringing a superiority of fire to bear on the most advanced portion of the skirmishers, unless the assailant presents a front almost equal to the defender. And since, when an attack is once pronounced enough to allow a defender to know what the special point to be assailed is, he will combine every means of defence he can to repel the assault at that one point, and will not fear to render his defence of other points less active, it is clear that nearly the whole fire of all neighbouring parts of a position will be brought in to defend the one assailed point. The increased length of range of the weapon will tell all the more in favour of the defender, because at such times he will not scruple to use it to the utmost. It will be no waste of ammunition in these cases to pour very long-range fire from neighbouring heights upon the enemy's skirmishers, as they begin to thicken at a particular point preparatory to the final rush. To put the case another way: Now that fire rather than assault is the great means by which to carry a position, any difference in the extent of the front of fire has become a much more important element than formerly. (See diagram 2.)

Is it not, then, a legitimate deduction, that it will be necessary to occupy an enemy's attention along almost the whole line, not merely by false assaults at particular points, but by such a general advance of skirmishers as shall force in all skirmishers from the front of the position? And though it will be essential not to commit any other part of the attacking line too much, will it not be necessary to make the general attack sufficiently vigorous to prevent the defender's fire from other points from being employed in the defence of the specially assailed part of the position?

From the time when the artillery can no longer support by their fire the assaulting columns, it will be very important for them to direct their action against all that part of the line which might in any way assist in opposing the attack. In so far as time permits, intrenchments should be thrown up and the ground prepared for defence throughout all that part of the assailant's position which is only intended to occupy the enemy and keep him from making a counter-assault. Shelter-trenches should generally be placed out of range of the defender's artillery at those points where the main reserves are kept. The epaulments for the artillery, however, must necessarily be in advance of these. It would probably be well to have some kind of shelter-trench for escorts near to the epaulments. The object, however, of these two classes of intrenchment would be essentially different. The first kind—that out of range of the enemy's artillery—is defensive, and should serve as a final rallying-point in case of local disaster; that for the artillery and their escorts, on the other hand, is intended to enable the guns to act effectively on the offensive. As much of the former work would be out of reach of danger, it would probably be possible in a friendly country to employ labourers to construct some of the works, under the direction of the engineers, in order not to fatigue the men. This, however, will probably be impossible in regard to the necessarily much exposed artillery intrenchments.

Are we to assume that always attack must be directed against a flank in the manner in which it was in most cases by the Germans during the late campaign? All recent writers on the subject appear to assume that it will be. It must be observed, moreover, that this phenomenon of constant flank attacks was quite as marked a feature of the American war as of the late one. The Americans seem to have habitually intrenched themselves against one another in front, and then to have worked round to a flank.* This was

* Chesney's 'Campaigns in Virginia.'

the use which was constantly made by Lee of Jackson's force. Probably, therefore, it may be assumed that the severity of fire along the front of a position will cause the first attempts to be nearly always made against one or other flanks. Obviously, however, as soon as this comes to be a recognised principle, and the assailing body begins to extend round to a flank, the defender will extend against him, and the problem will be to know exactly at what point the possible degree of extension is reached and passed. It will therefore require all a general's care to know how far to trust the defensive power of his own line, and when his enemy has overstrained his. At present we have no data on which to found a judgment as to what the defensive power of the breech-loader, in cases of extreme extension, is likely to be against very good troops. It is a question which it will take all the genius of the greatest generals to solve in each particular case. The elements concerned in the matter are obviously of infinite variety; the moral are probably more important than the material.*

---

* It may be sufficient to point out generally that the detail of *retreat* from a position, whether on the offence or on the defence, does not fall within the terms of the proposed subject. It is necessary, however, to the completeness of the argument in favour of the proposed general scheme of fighting on either side, to observe that the risks of retreat are not enhanced by attempting to meet the difficulties of our present form of fighting by organisation and successive reserves, rather than by in vain straining to maintain a close-order fight which inevitably dissolves in practice. The distinction between the form in which the Prussians and French severally fought after each had begun to realise the necessity for change, was not that the French were less scattered than the Prussians; on the contrary, they are expressly said to have been much more so. The distinction was this, the Prussian training had prepared them to be in hand though scattered. The French, unprepared for any other use of skirmishes than an auxiliary one, were out of hand as soon as they were scattered. The detail of retreat has been suggested by Prince Hohenlohe.

THE DEFENCE.

(*d.*) MODE OF COMBINING AND EMPLOYING THE DIFFERENT ARMS FOR RECEIVING THE ATTACK OF AN ENEMY.

THE defensive, like the offensive, appears chiefly to require such modifications as will give the weapons fuller play, and will as far as possible diminish the effect of the extreme destruction of the enemy's fire. Under this aspect, some rules which have passed down to us from times when the effect of fire was less the all-governing consideration than it is now, seem to require modification. It will almost always be infinitely better for the defender to select such a position as will enable him to bring a thoroughly sweeping fire to bear on all points, than to defend ground which presents physical difficulties to an enemy's approach, but affords the latter some degree of cover from fire. This is now true, even in a case in which the defender is restricted entirely to the defensive. The objections which have always existed to having in front of a defensive position much difficult ground, retain all their force in the usual case of a defensive preparatory to offence. Perhaps, however, sometimes the most difficult ground of all over which to pass from defence to offence will now be that which is perfectly open, but on the further side of which an enemy can secure a satisfactory position. It has been stated on good authority, though not, so far as I am aware, in any of the histories of the war, that at Spicheren the French skirmishers were unable to remain in advance of the main body, because, in consequence of the extreme steepness of the ground in front of the position, they could not have made any safe retreat in presence of the terrible fire of those advancing against them. This in itself suggests a new objection to the assumption of positions dependent for their strength on the steepness of the fall of the ground. In almost all cases where a position has fallen during the late war, except when the defender's ammunition has been exhausted, the approach has been made over difficult ground not well exposed to fire.

The first thing, therefore, to be obtained, if it can be in any way secured, even at a sacrifice of more perfect cover, is as clear a view and field for fire as may be; not less than 2500 yards if possible.*

I cannot help thinking that we are too apt to underrate the importance of this in our preparations for defence. Nothing can be more admirable than the principles advocated in our field-exercise book on the subject of rapid intrenchment. But for one place

---

* Das heutige Gefecht, p. 29.

that has fallen, because the men on the defensive had not sufficient cover, a dozen have fallen because the obstacles to fire in the front have not been cleared away. The very fact that men are on the defensive, and therefore stationary, gives them, if they use the natural cover obtainable, so much less exposure than those who must move rapidly from position to position, that it ought to be the object of the defence to emphasise this advantage which they already possess, by depriving the assailants of all chance of finding shelter. It would seem, therefore, quite as important to train men habitually to calculate how much ground can be cleared in a given time, and in what way it can best be done, as to teach them how to throw up intrenchments. Moreover, what were obstacles advantageous to the defendant in former times have ceased to be so. In principle, no doubt, an obstacle running perpendicularly to the front of a position, and ending there, will now as ever divide an assailant's forces, and therefore be of advantage to the defender. But it seems no longer to be true that this applies to hedgerows and minor obstacles of that kind. Formerly a bayonet-charge might have driven back the enemy along one side of a hedge, without its being possible that he should be supported from the other side. The risk might then have been serious. Now the support against the counter-assault afforded by skirmishers moving up to the hedge and firing through it, would be almost as great as if the hedge were not there.

From the crossing musketry-fire of the main position the assailant is to no inconsiderable extent protected by such obstacles as these. Hence the more of a clear free-swept glacis-slope the defendant can obtain the better. At Gravelotte* the Prussians climbed successfully the steep hills with their well-wooded sides, only to find all effort to pass out on to the smooth glacis beyond merely lead to their being massacred. I cannot think that any one who has once seen those Gravelotte graves would ever again doubt what is the sort of position to prepare for an enemy.

I do not, however, in the least mean to undervalue the importance of preparing an intrenched position, or that of placing effective obstacles in an adversary's way. But we realise the protection afforded by the one far more than that secured by the other. The principle urged by Boguslawski seems thoroughly sound, that on the defensive you should train your men to understand, "if you don't go away, the enemy will." The moral effect, to this end, of accustoming the men to see every possible obstacle that could give cover to the enemy cleared off, is sure to be very great. They will feel that the only thing they want is, that the enemy should show himself in order that they may destroy him. It will directly tend to increase that confidence of the man in his weapon,

---

* I speak here only of the character of the front of the position at Gravelotte. In other respects it was very defective.

which is so essential a part now of the preparation of a soldier for fighting effectively.*

The next most important point—and it is, perhaps, an open question whether it ought not to be considered the more important of the two—is, that the position should have sufficient manœuvring depth. The space at present required is far greater than it ever has been; for formations should be very deep, both in order, to some extent, to secure proper cover for the reserves, and, above all, in order to meet the constant efforts of the enemy to attack a flank. Special cover for the supports and reserves, as near to the front as possible, is, however, an immense advantage.

A smaller number of men in proportion than formerly have usually been placed by the Prussians in the front line. The general principles adopted now are, however, very nearly those which were advocated by Count von Moltke in '65.† They consist in placing a very small proportion of infantry—say one brigade per corps d'armée—to occupy the ground subsequently to be held in force. With these a considerable number of guns are, however, provided at the most commanding positions. No attempt is made to occupy actually the whole of the position; but a second body—say another brigade—is told off ex-

---

* Boguslawski, p. 168.
† Von Moltke, with clear foresight of events, p. 17; Das heutige Gefecht, p. 20; Boguslawski, p. 151, 167, &c.

pressly to support the first, the remainder of the corps being kept in hand ready for all emergencies, in order to extend the position if the enemy attempts to turn it, or to support the front if the attack be a direct one.

The flanks should be protected either by natural obstacles or by strong defensive positions with a good command over the whole country round. Where an army is absolutely restricted to the defensive, the natural obstacle may be the best. It is probably on this ground that the Duke of Wurtemberg maintains that at Gravelotte the French ought to have extended to the river Orne. The risks of front attack being so serious, an enemy who is prevented by a natural obstacle on either flank from outflanking has certainly a difficult task before him. But for an enemy on the defensive, which intends to be ready to assume the offensive if opportunity offers, it is difficult to see that the increased power of the weapons has made it less advantageous than it was formerly to rest on a strongly defensible flank on which it may be useful to pivot. Even in the case as it stood, the Prussians did not take St Privat because they succeeded in passing into the ground between the Orne and that village, but because the French had absolutely exhausted all their ammunition.

In preparing the defence every effort should be made to select one strong line on ground which is to be held at all hazards, and not a series, since both the destruction and the demoralising effect of retreat is so great under the awful effects of modern fire. If very strong advance posts, well connected with the main position, can be secured, they will be of immense service, especially in interfering with the development of the enemy's artillery-attack on the main position, and in preventing the enemy from ascertaining exactly what the nature of the position is.* If the advance posts, however, would be weak, it would be much better to dispense with them altogether, because their fall discourages the men, and there is difficulty in preventing the enemy from approaching the main position more easily if he follows closely their defenders. In case no advance posts are thrown out, posts of observation should be established, and held as long as possible; but no attempt should be made to defend them.

For the protection of the flanks, supports may safely be placed in echelon behind them, since fire is now so deadly that the enemy gains nothing by getting round the head of the flank if it be supported by fresh force echeloned behind it.† Indeed, a series of false flanks or concealed echelons would seem to be the best way of meeting the attempt to outflank. As far as possible, a position should be selected which will enable the first supports to be close up to the skirmishers, who will line the immediate front.

As soon as the enemy's attack has fully developed

* Boguslawski, p. 167.   † Ibid.

itself, and the point he is intending to assault is known, it is better to trust the direct defence of the position to fresh clouds of infantry skirmishers pushed up to it as the enemy approaches.* The artillery should be kept back just in rear of the position, to open as soon as the enemy fairly gets in. It should prepare the way for a powerful counter-attack. Fresh infantry and cavalry ought to be behind the guns, to assault on the enemy's first fairly arriving, whilst he is still finding difficulty in getting cover in a position he does not know. Whether on the defensive on the whole, or in that part of an assailant's position which is restricted to the defensive, great as are the advantages of maintaining a local defence as long as possible, the moral effect of attack must never be forgotten. It is an element that must be freely employed, whether to impose on an enemy, as was so brilliantly illustrated by Alvensleben when he was clinging with weak forces to fiercely-assailed positions at Rezonville, or in order to revive the drooping spirits of discouraged men.

Assuming, however, that the part attempted will be, in the first instance, the flank of the whole position, the right tactics for the defender would seem to be to deceive the enemy, by every means in his power, as to the place on which his flank rests; and having succeeded in committing him, if possible, to

* German General, p. 249; Boguslawski as above.

an excessive extension, to strike at the point of greatest weakness in the general line opposed to him.

One other remark must be added. Now, as formerly, the tactics of defence differ essentially from those of offence in this. A certain liberty of action for subordinates has always been necessary in taking the offensive. That liberty of action must now be very largely increased. But on the defensive it has been always essential for the general to have nearly the whole of his forces perfectly under immediate command. Circumstances have partly modified this, no doubt, since it seems essential, according to the evidence we have, that the means of counter-attack, even within a position, should be chiefly by skirmishing lines; but the main fact remains as before. For this reason, then, if for no other, until a greater manœuvring facility—due as much to organisation as to training—has been acquired by our army, the defensive would seem to be the *rôle* we ought to seek. Unhappily, no army can limit itself to the defensive. Even on the defensive, all that now remains in the power of the commander is to determine the moment at which he shall abandon his absolute dictation, and trust, as he launches his troops into counter-attack, to their readiness to conform, and their capacity for conforming, to the essence of his instructions. For this, therefore, whether in offence or in defence, it is inevitable that we should now be ready to prepare our army. If we cannot do so, the best system of manœuvres may be a subject of interesting discussion in the abstract for a military student; it can never be a means of victory for an army, or of safety for a nation.

THE END.

DIAGRAM I.

FORM OF ATTACK OF BRIGADE. IN FIRST LINE.

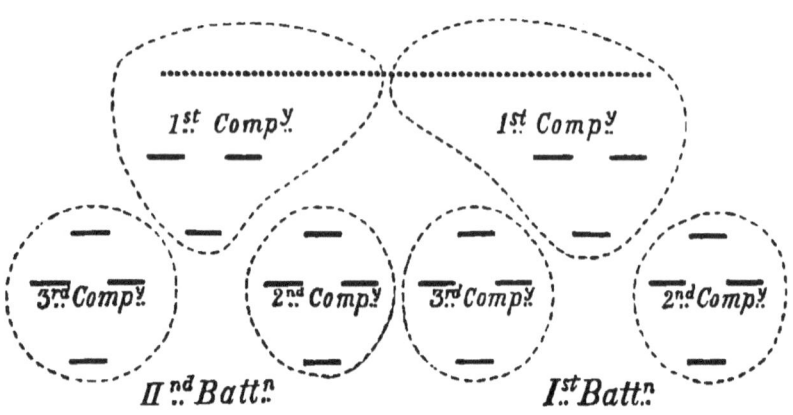

DIAGRAM II.

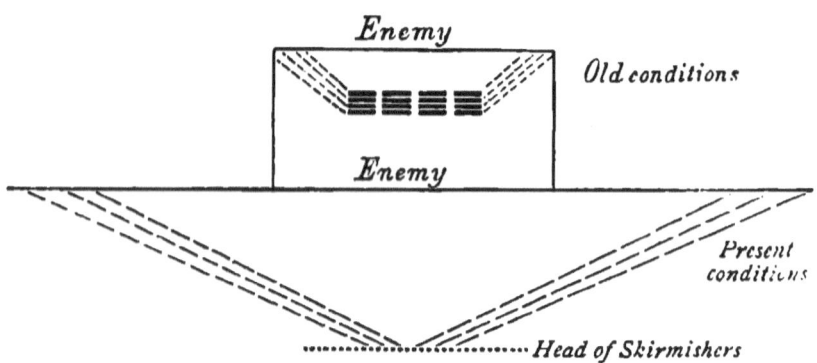

Showing greater length of Infantry line which would be available against a modern local assault for the two reasons assigned. As soon as the Artillery on the offensive is silenced by the advance of the offensive Infantry, the same illustration would apply with even greater force to the increased front of defensive Artillery, which would be able to play on an advancing local attack.

LATELY PUBLISHED.

THE OPERATIONS OF WAR Explained and Illustrated. By EDWARD BRUCE HAMLEY, Colonel in the Army, and Lieut.-Colonel Royal Artillery; Commandant Staff-College. Second Edition. Revised throughout by the Author, and containing important additions on the influence of Railways and Telegraphs on War, and on the effects which the changes in Weapons may be expected to produce in Tactics. Quarto, 17 Maps and Plans, with other Illustrations, £1, 8s.

BY SAME AUTHOR,

THE STORY OF THE CAMPAIGN OF SEBASTOPOL. Written in the Camp. With Illustrations by the Author. 8vo, 21s.

WELLINGTON'S CAREER: A Military and Political Summary. Crown 8vo, 2s.

THE WAR FOR THE RHINE FRONTIER, 1870: Its Political and Military History. By COL. W. RÜSTOW. Translated from the German by JOHN LAYLAND NEEDHAM, Lieutenant R.M. Artillery. 3 vols. 8vo, with Maps and Plans, £1, 11s. 6d.

JOURNAL OF THE WATERLOO CAMPAIGN. Kept throughout the Campaign of 1815. By GENERAL CAVALIÉ MERCER, Commanding the 9th Brigade Royal Artillery. 2 vols. post 8vo, 21s.

REMINISCENCES OF A VOLUNTEER.

THE BATTLE OF DORKING. From Seventh Edition 'Blackwood's Magazine' for May. Second Hundredth Thousand. 6d.

THE HANDY HORSE-BOOK; or, Practical Instructions in Driving, Riding, and the General Care and Management of Horses. By MAGENTA. Fifth Edition, with Illustrations. Crown 8vo, 4s. 6d.

SIR ARCHIBALD ALISON'S HISTORY OF EUROPE, FROM THE FRENCH REVOLUTION TO THE ACCESSION OF LOUIS NAPOLEON.

1st Series, 1789-1815, 14 vols. 8vo, £10, 10s. People's Edition, 12 vols., 51s.

2d Series, 1815-1852, 9 vols. 8vo, £6, 7s. 6d. People's Edition, 8 vols., 34s.

ATLAS ILLUSTRATIVE OF THE HISTORY OF EUROPE DURING THE FRENCH REVOLUTION, containing 109 Maps and Plans of Countries, Battles, Sieges, and Sea-Fights, constructed by A. KEITH JOHNSTON, LL.D. F.R.S.E., &c., with a Vocabulary of Military and Marine Terms. Royal Quarto, £3, 3s. Another Edition, £1, 11s. 6d.

THE INVASION OF THE CRIMEA. By ALEXANDER WILLIAM KINGLAKE, M.P. Four vols. 8vo, with Maps and Plans, £3, 6s.

BIOGRAPHIES OF EMINENT SOLDIERS OF THE LAST FOUR CENTURIES. By MAJOR-GENERAL JOHN MITCHELL, Author of 'Life of Wallenstein,' 'The Fall of Napoleon,' &c. Edited, with a Memoir of the Author, by LEONHARD SCHMITZ, LL.D. 8vo, 9s.

THE SUBALTERN. By G. R. GLEIG, Chaplain-General of Her Majesty's Forces. Library Edition, Revised and Corrected, with a New Preface. Crown 8vo, 7s. 6d.

PROFESSOR AYTOUN'S WORKS.

LAYS OF THE SCOTTISH CAVALIERS. An Illustrated Edition. From Designs by Sir JOSEPH NOEL PATON. 21s. ANOTHER EDITION, the 22d, fcap 8vo, 7s. 6d.

BOTHWELL: A Poem. Third Edition. Fcap. 8vo, 7s. 6d.

THE BALLADS OF SCOTLAND. Edited by Professor AYTOUN. Fourth Edition. 2 vols. fcap. 8vo, 12s.

THE BOOK OF BALLADS. Edited by BON GAULTIER. With Illustrations by DOYLE, LEECH, and CROWQUILL. Eleventh Edition, cloth, gilt edges, 8s. 6d.

SONGS AND VERSES. SOCIAL AND SCIENTIFIC. By an OLD CONTRIBUTOR TO 'MAGA.' A New Edition. Fcap. 8vo, 3s. 6d. With Music of some of the Songs.

WILLIAM BLACKWOOD & SONS, Edinburgh and London.

EDUCATIONAL WORKS

PUBLISHED BY

WILLIAM BLACKWOOD & SONS,

EDINBURGH AND LONDON.

Zoology.

A MANUAL OF ZOOLOGY, FOR THE USE OF STUDENTS. With a General Introduction on the Principles of Zoology. By HENRY ALLEYNE NICHOLSON, M.D., F.R.S.E., F.G.S., &c., Professor of Natural History in the University of Toronto. Second Edition. Crown 8vo, pp. 674, with 243 Engravings on Wood, 12s. 6d.

"It is the best manual of zoology yet published, not merely in England, but in Europe."—*Pall Mall Gazette, July* 20, 1871.
"The best treatise on Zoology in moderate compass that we possess."—*Lancet, May* 18, 1872.

BY THE SAME AUTHOR.

TEXT-BOOK OF ZOOLOGY, FOR THE USE OF SCHOOLS. Crown 8vo, with numerous Engravings on Wood, 6s.

"This capital introduction to natural history is illustrated and well got up in every way. We should be glad to see it generally used in schools."—*Medical Press and Circular.*

BY THE SAME AUTHOR.

INTRODUCTORY TEXT-BOOK OF ZOOLOGY, FOR THE USE OF JUNIOR CLASSES. With 127 Engravings, 3s. 6d.

"Very suitable for junior classes in schools. There is no reason why any one should not become acquainted with the principles of the science, and the facts on which they are based, as set forth in this volume."—*Lancet.*

BY THE SAME AUTHOR.

INTRODUCTION TO THE STUDY OF BIOLOGY. Crown 8vo, with numerous Engravings, 5s.

"Admirably written and fairly illustrated, and brings within the compass of 160 pages the record of investigations and discoveries scattered over as many volumes. Seldom indeed do we find such subjects treated in a style at once so popular and yet so minutely accurate in scientific detail."—*Scotsman.*

BY THE SAME AUTHOR—In the Press.

A MANUAL OF PALÆONTOLOGY, FOR THE USE OF STUDENTS. Crown 8vo, with upwards of 400 Engravings.

Botany.

ADVANCED TEXT-BOOK OF BOTANY. For the Use of Students. By ROBERT BROWN, M.A., PH.D., F.R.G.S., Lecturer on Botany under the Science and Art Department of the Committee of the Privy Council on Education.
[*In the press.*

Geology.

INTRODUCTORY TEXT-BOOK OF GEOLOGY. By David Page, LL.D., Professor of Geology in the Durham University of Physical Science, Newcastle. With Engravings on Wood and Glossarial Index. Ninth Edition. 2s.

ADVANCED TEXT-BOOK OF GEOLOGY, Descriptive and Industrial. By the Same. With Engravings, and Glossary of Scientific Terms. Fifth Edition, revised and Enlarged. 7s. 6d.

"We have carefully read this truly satisfactory book, and do not hesitate to say that it is an excellent compendium of the great facts of Geology, and written in a truthful and philosophic spirit."—*Edinburgh Philosophical Journal.*

"As a school-book nothing can match the Advanced Text-Book of Geology by Professor Page of Newcastle."—*Mechanic's Magazine.*

THE GEOLOGICAL EXAMINATOR. A Progressive Series of Questions, adapted to the Introductory and Advanced Text-Books of Geology. Prepared to assist Teachers in framing their Examinations, and Students in testing their own Progress and Proficiency. By the Same. Fourth Edition. 9d.

GEOLOGY FOR GENERAL READERS. A Series of Popular Sketches in Geology and Palæontology. By the Same. Third Edition, enlarged. 6s.

"This is one of the best of Mr Page's many good books. It is written in a flowing popular style. Without illustration or any extraneous aid, the narrative must prove attractive to any intelligent reader."—*Geological Magazine.*

HANDBOOK OF GEOLOGICAL TERMS, GEOLOGY, AND PHYSICAL GEOGRAPHY. By the Same. Second Edition, enlarged. 7s. 6d.

CHIPS AND CHAPTERS. A Book for Amateurs and Young Geologists. By the Same. 5s.

THE PAST AND PRESENT LIFE OF THE GLOBE. With numerous Illustrations. By the Same. Crown 8vo. 6s.

THE CRUST OF THE EARTH: A Handy Outline of Geology. By the Same. 1s.

"An eminently satisfactory work, giving, in less than 100 pages, an admirable outline sketch of Geology, . . . forming, if not a royal road, at least one of the smoothest we possess to an intelligent acquaintance with geological phenomena."—*Scotsman.*

"Of singular merit for its clearness and trustworthy character."—*Standard.*

"*Few of our handbooks of popular science can be said to have greater or more decisive merit than those of Mr Page on Geology and Palæontology. They are clear and vigorous in style, they never oppress the reader with a pedantic display of learning, nor overwhelm him with a pompous and superfluous terminology; and they have the happy art of taking him straightway to the face of nature herself, instead of leading him by the tortuous and bewildering paths of technical system and artificial classification.*"—Saturday Review.

Meteorology.

INTRODUCTORY TEXT-BOOK OF METEOROLOGY. By ALEXANDER BUCHAN, M.A., F.R.S.E., Secretary of the Scottish Meteorological Society, Author of 'Handy Book of Meteorology,' &c. Crown 8vo, with 8 Coloured Charts and other Engravings, pp. 218. 4s. 6d.

"A handy compendium of Meteorology by one of the most competent authorities on this branch of science."—*Petermanns Geographische Mittheilungen.*

"We can recommend it as a handy, clear, and scientific introduction to the theory of Meteorology, written by a man who has evidently mastered his subject."—*Lancet.*

"An exceedingly useful volume."—*Athenæum.*

History.

EPITOME OF ALISON'S HISTORY OF EUROPE, FOR THE USE OF SCHOOLS. Sixteenth Edition. Post 8vo, pp. 604. 7s. 6d. bound in leather.

ATLAS TO EPITOME OF THE HISTORY OF EUROPE. ELEVEN COLOURED MAPS. By A. KEITH JOHNSTON, LL.D., F.R.S.E. In 4to, 7s.

THE EIGHTEEN CHRISTIAN CENTURIES. By the Rev. JAMES WHITE, Author of 'The History of France.' Seventh Edition, post 8vo, with Index, 6s.

"He has seized the salient points—indeed, the governing incidents—in each century, and shown their received bearing as well on their own age as on the progress of the world. Vigorously and briefly, often by a single touch, has he marked the traits of leading men; when needful, he touches slightly their biographical career. The state of the country and of society, of arts and learning, and, more than all, of the modes of living, are graphically sketched, and, upon the whole, with more fulness than any other division."—*Spectator.*

HISTORY OF FRANCE, FROM THE EARLIEST TIMES. By the Rev. JAMES WHITE, Author of 'The Eighteen Christian Centuries.' Fifth Edition, post 8vo, with Index, 6s.

FACTS AND DATES; or, The Leading Events in Sacred and Profane History, and the Principal Facts in the Various Physical Sciences: the Memory being aided throughout by a Simple and Natural Method. For Schools and Private Reference. By the Rev. ALEX. MACKAY, LL.D., F.R.G.S., Author of 'A Manual of Modern Geography,' &c. Second Edition, crown 8vo, pp. 336. 4s.

"A most valuable book of reference, which will be of immense service to students of history. His wide knowledge has directed the author at once to the most trustworthy guides in the various departments of the almost illimitable field he has traversed. . . . Every date throughout is embodied in a mnemonic sentence, so happily and tersely illustrative of the event, as to leave us fairly astonished at the patience and ingenuity of the author."—*Papers for the Schoolmaster.*

Geography.

NEW AND GREATLY IMPROVED EDITION.
A MANUAL OF MODERN GEOGRAPHY, MATHEMATICAL, PHYSICAL, AND POLITICAL. By the Rev. ALEXANDER MACKAY, LL.D., F.R.G.S. Crown 8vo, pp. 676. 7s. 6d.

This volume—the result of many years' unremitting application—is specially adapted for the use of Teachers, Advanced Classes, Candidates for the Civil Service, and proficients in geography generally.

TWENTIETH THOUSAND.
ELEMENTS OF MODERN GEOGRAPHY. By the Same. Eleventh Edition, revised to the present time. Crown 8vo, pp. 300. 3s.

The 'Elements' form a careful condensation of the 'Manual,' the order of arrangement being the same, the river-systems of the globe playing the same conspicuous part, the pronunciation being given, and the results of the latest census being uniformly exhibited. This volume is now extensively introduced into many of the best schools in the kingdom.

FIFTIETH THOUSAND.
OUTLINES OF MODERN GEOGRAPHY: TWELFTH EDITION, REVISED TO THE PRESENT TIME. By the Same. 18mo, pp. 112. 1s.

These 'Outlines'—in many respects an epitome of the 'Elements'—are carefully prepared to meet the wants of beginners. The arrangement is the same as in the Author's larger works. Minute details are avoided, the broad outlines are graphically presented, the accentuation marked, and the most recent changes in political geography exhibited.

THIRTY-EIGHTH THOUSAND, REVISED TO THE PRESENT TIME.
FIRST STEPS IN GEOGRAPHY. By the Same. 18mo, pp. 56. Sewed, 4d. In cloth, 6d.

GEOGRAPHY OF THE BRITISH EMPIRE. From 'First Steps in Geography.' By the Same. 3d.

Physical Geography.

INTRODUCTORY TEXT-BOOK OF PHYSICAL GEOGRAPHY. With Sketch-Maps and Illustrations. By DAVID PAGE, LL.D., Professor of Geology in the Durham University of Physical Science, Newcastle. Fifth Edition. 2s.

ADVANCED TEXT-BOOK OF PHYSICAL GEOGRAPHY. By the Same. With Engravings. 5s.
"A thoroughly good Text-Book of Physical Geography."—*Saturday Review.*

EXAMINATIONS ON PHYSICAL GEOGRAPHY. A Progressive Series of Questions, adapted to the Introductory and Advanced Text-Books of Physical Geography. By the Same. 9d.

COMPARATIVE GEOGRAPHY. By CARL RITTER. Translated by W. L. GAGE. Fcap., 3s. 6d.

Geographical Class-Books.

OPINIONS OF DR MACKAY'S SERIES.

MANUAL.

Annual Address of the President of the Royal Geographical Society (Sir Roderick I. Murchison). We must admire the ability and persevering research with which he has succeeded in imparting to his Manual so much freshness and originality. In no respect is this character more apparent than in the plan of arrangement, by which the author commences his description of the physical geography of each tract by a sketch of its true basis or geological structure. The work is largely sold in Scotland, but has not been sufficiently spoken of in England. It is, indeed, a most useful school-book in opening out geographical knowledge.

Saturday Review.—It contains a prodigious array of geographical facts, and will be found useful for reference.

English Journal of Education.—Of all the Manuals on Geography that have come under our notice, we place the one whose title is given above in the first rank. For fulness of information, for knowledge of method in arrangement, for the manner in which the details are handled, we know of no work that can, in these respects, compete with Mr Mackay's Manual.

ELEMENTS.

A. KEITH JOHNSTON, LL.D., F.R.S.E., F.R.G.S., H.M. Geographer for Scotland, Author of the 'Physical Atlas,' &c. &c.—There is no work of the kind in this or any other language, known to me, which comes so near my *ideal* of perfection in a school-book, on the important subject of which it treats. In arrangement, style, selection of matter, clearness, and thorough accuracy of statement, it is without a rival; and knowing, as I do, the vast amount of labour and research you bestowed on its production, I trust it will be so appreciated as to insure, by an extensive sale, a well-merited reward.

G. BICKERTON, Esq., Edinburgh Institution.—I have been led to form a very high opinion of Mackay's 'Manual of Geography' and 'Elements of Geography,' partly from a careful examination of them, and partly from my experience of the latter as a text-book in the EDINBURGH INSTITUTION. One of their most valuable features is the elaborate Table of River-Basins and Towns, which is given in addition to the ordinary Province or County list, so that a good idea may be obtained by the pupil of the natural as well as the political relationship of the towns in each country. On all matters connected with Physical Geography, Ethnography, Government, &c., the information is full, accurate, and well digested. They are books that can be strongly recommended to the student of geography.

RICHARD D. GRAHAM, English Master, College for Daughters of Ministers of the Church of Scotland and of Professors in the Scottish Universities.—No work with which I am acquainted so amply fulfils the conditions of a perfect text-book on the important subject of which it treats, as Dr Mackay's 'Elements of Modern Geography.' In fulness and accuracy of details, in the scientific grouping of facts, combined with clearness and simplicity of statement, it stands alone, and leaves almost nothing to be desired in the way of improvement. Eminently fitted, by reason of this exceptional variety and thoroughness, to meet all the requirements of higher education, it is never without a living interest, which adapts it to the intelligence of ordinary pupils. It is not the least of its merits that its information is abreast of all the latest developments in geographical science, accurately exhibiting both the recent political and territorial changes in Europe, and the many important results of modern travel and research.

Spectator.—The best Geography we have ever met with.

IMPROVED EDITIONS.

School Atlases.
By A. KEITH JOHNSTON, LL.D., &c.
Author of the Royal and the Physical Atlases, &c.

ATLAS OF GENERAL AND DESCRIPTIVE GEO-
GRAPHY. A New and Enlarged Edition, suited to the best Text-Books; with Geographical information brought up to the time of publication. 26 Maps, clearly and uniformly printed in colours, with Index. Imp. 8vo. Half-bound, 12s. 6d.

ATLAS OF PHYSICAL GEOGRAPHY, illustrating, in a Series of Original Designs, the Elementary Facts of GEOLOGY, HYDROGRAPHY, METEOROLOGY, and NATURAL HISTORY. A New and Enlarged Edition, containing 4 new Maps and Letterpress. 20 Coloured Maps. Imp. 8vo. Half-bound, 12s. 6d.

ATLAS OF ASTRONOMY. A New and Enlarged Edition, 21 Coloured Plates. With an Elementary Survey of the Heavens, designed as an accompaniment to this Atlas, by ROBERT GRANT, LL.D., &c., Professor of Astronomy and Director of the Observatory in the University of Glasgow. Imp. 8vo. Half-bound, 12s. 6d.

ATLAS OF CLASSICAL GEOGRAPHY. A New and Enlarged Edition. Constructed from the best materials, and embodying the results of the most recent investigations, accompanied by a complete INDEX OF PLACES, in which the proper quantities are given by T. HARVEY and E. WORSLEY, MM.A. Oxon. 21 Coloured Maps. Imp. 8vo. Half-bound, 12s. 6d.

"This Edition is so much enlarged and improved as to be virtually a new work, surpassing everything else of the kind extant, both in utility and beauty." —*Athenæum*.

ELEMENTARY ATLAS OF GENERAL AND
DESCRIPTIVE GEOGRAPHY, for the Use of Junior Classes; including a MAP OF CANAAN and PALESTINE, with GENERAL INDEX. 8vo, half-bound, 5s.

NEW ATLAS FOR PUPIL-TEACHERS.

THE HANDY ROYAL ATLAS. 45 Maps clearly printed and carefully coloured, with GENERAL INDEX. Imp. 4to, £2, 12s. 6d., half-bound morocco.

This work has been constructed for the purpose of placing in the hands of the public a useful and thoroughly accurate ATLAS of Maps of Modern Geography, in a convenient form, and at a moderate price. It is based on the 'ROYAL ATLAS,' by the same Author; and, in so far as the scale permits, it comprises many of the excellences which its prototype is acknowledged to possess. The aim has been to make the book strictly what its name implies, a HANDY ATLAS—a valuable substitute for the 'Royal,' where that is too bulky or too expensive to find a place, a needful auxiliary to the junior branches of families, and a *vade mecum* to the tutor and the pupil-teacher.

Arithmetic, &c.

THE THEORY OF ARITHMETIC. By DAVID MUNN, F.R.S.E., Mathematical Master, Royal High School of Edinburgh. Crown 8vo, pp. 294. 5s.

"We want books of this kind very much—books which aim at developing the educational value of Arithmetic by showing how admirably it is calculated to exercise the thinking powers of the young. Your book is, I think, excellent —brief but clear; and I look forward to the good effects which it shall produce, in awaking the minds of many who regard Arithmetic as a mere mechanical process."—*Professor Kelland.*

ELEMENTARY ARITHMETIC. By EDWARD SANG, F.R.S.E. This Treatise is intended to supply the great desideratum of an intellectual instead of a routine course of instruction in Arithmetic. Post 8vo, 5s.

THE HIGHER ARITHMETIC. By the same Author. Being a Sequel to 'Elementary Arithmetic.' Crown 8vo, 5s.

FIVE-PLACE LOGARITHMS. Arranged by E. SANG, F.R.S.E. Sixpence. For the Waistcoat-Pocket.

TREATISE ON ARITHMETIC, with numerous Exercises for Teaching in Classes. By JAMES WATSON, one of the Masters of Heriot's Hospital. Foolscap, 1s.

EXAMPLES IN BOOK-KEEPING; A TREATISE SPECIALLY ADAPTED FOR SCHOOLS. By the Rev. JOHN CONSTABLE, M.A., Principal of the Royal Agricultural College, Cirencester. [*In the press.*

A GLOSSARY OF NAVIGATION. Containing the Definitions and Propositions of the Science, Explanation of Terms, and Description of Instruments. By the Rev. J. B. HARBORD, M.A., Assistant Director of Education, Admiralty. Crown 8vo, Illustrated with Diagrams, 6s.

DEFINITIONS AND DIAGRAMS IN ASTRONOMY AND NAVIGATION. By the Same. 1s. 6d.

ELEMENTARY HANDBOOK OF PHYSICS. With 210 Diagrams. By WILLIAM ROSSITER, F.R.A.S., &c. Crown 8vo, pp. 390. 5s.

"A singularly interesting Treatise on Physics, founded on facts and phenomena gained at first hand by the Author, and expounded in a style which is a model of that simplicity and ease in writing which betokens mastery of the subject. To those who require a non-mathematical exposition of the principles of Physics a better book cannot be recommended."—*Pall Mall Gazette.*

ON PRIMARY INSTRUCTION IN RELATION TO EDUCATION. By SIMON S. LAURIE, A.M. Author of 'Philosophy of Ethics,' &c. Crown 8vo, 4s. 6d.

English Language.

ETYMOLOGICAL AND PRONOUNCING DICTIONARY OF THE ENGLISH LANGUAGE. Including a very Copious Selection of Scientific Terms. For use in Schools and Colleges, and as a Book of General Reference. By the Rev. JAMES STORMONTH. The Pronunciation carefully Revised by the Rev. P. H. PHELP, M.A. Cantab. Crown 8vo, pp. 755. 7s. 6d.

BY THE SAME AUTHOR.

THE SCHOOL ETYMOLOGICAL DICTIONARY AND WORD-BOOK. Combining the advantages of an ordinary School Dictionary and an Etymological Spelling-Book. Fcap. 8vo, pp. 220. [*In the press.*

PROGRESSIVE AND CLASSIFIED SPELLING-BOOK. By HANNAH R. LOCKWOOD, Authoress of 'Little Mary's Mythology.' Fcap. 8vo, 1s. 6d.

A MANUAL OF ENGLISH PROSE LITERATURE, Biographical and Critical: designed mainly to show characteristics of style. By W. MINTO, M.A. Crown 8vo, 10s. 6d.

"Is a work which all who desire to make a close study of style in English prose will do well to use attentively."—*Standard.*

"It is not often, among the innumerable manuals on the same subject, that it is our good fortune to come across so excellent a book as that before us. Carefully and thoughtfully written, it indeed offers a very marked contrast to the shallow and hastily-compiled productions that crowd our book market, while its originality of conception and manner makes it specially valuable when so many of the compilers of similar works seem to go to one and the same source not only for their facts but for their conclusions."—*Educational Times.*

"Here we do not find the *crambe repetita* of old critical formulæ, the simple echoes of superannuated rhetorical dicta, but a close and careful analysis of the main attributes of style, as developed in the work of its greatest masters, stated with remarkable clearness of expression, and arranged upon a plan of most exact method. Nothing can be well conceived more consummate as a matter of skill than the analytical processes of the writer as he lays bare to our view the whole anatomy—even every joint and sinew and artery in the framework —of the sentence he dissects, and as he points out their reciprocal relations, their minute interdependencies."—*School Board Chronicle.*

"An admirable book, well selected and well put together."—*Westminster Review.*

"In many of the sketches might be pointed out felicitous analyses of character, as well as acute and searching criticism; of all which, however, no extracts within the limits of a notice could give any adequate idea."—*Examiner.*

"Mr Minto's is no common book, but a very careful and well-considered survey of the wide field he traverses—a survey undertaken not without considerable competency and large equipment of knowledge."—*Scotsman.*

ENGLISH PROSE COMPOSITION: A Practical
Manual for Use in Schools. By James Currie, M.A., Principal of the Church of Scotland Training College, Edinburgh. Seventh Edition, 1s. 6d.

"We do not remember having seen a work so completely to our mind as this, which combines sound theory with judicious practice. Proceeding step by step, it advances from the formation of the shortest sentences to the composition of complete essays, the pupil being everywhere furnished with all needful assistance in the way of models and hints. Nobody can work through such a book as this without thoroughly understanding the structure of sentences, and acquiring facility in arranging and expressing his thoughts appropriately. It ought to be extensively used."—*Athenæum.*

A TREASURY OF THE ENGLISH AND GERMAN LANGUAGES.
Compiled from the best Authors and Lexicographers in both Languages. Adapted to the Use of Schools, Students, Travellers, and Men of Business; and forming a Companion to all German-English Dictionaries. By Joseph Cauvin, LL.D. & Ph.D., of the University of Göttingen, &c. Crown 8vo, 7s. 6d., bound in cloth.

"An excellent English-German Dictionary, which supplies a real want."—*Saturday Review.*

"The difficulty of translating English into German may be greatly alleviated by the use of this copious and excellent English-German Dictionary, which specifies the different senses of each English word, and gives suitable German equivalents. It also supplies an abundance of idiomatic phraseology, with many passages from Shakespeare and other authors aptly rendered in German. Compared with other dictionaries, it has decidedly the advantage."—*Athenæum.*

CHOIX DES MEILLEURES SCENES DE MOLIERE,
avec des Notes de Divers Commentateurs, et autres Notes Explicatives. Par Dr E. Dubuc. Fcap. 8vo, 4s. 6d.

Agriculture.

CATECHISM OF PRACTICAL AGRICULTURE.
By Henry Stephens, F.R.S.E., Author of the 'Book of the Farm.' With Engravings. 1s.

PROFESSOR JOHNSTON'S CATECHISM OF AGRICULTURAL CHEMISTRY.
A New Edition, edited by Professor Voelcker. With Engravings. 1s.

PROFESSOR JOHNSTON'S ELEMENTS OF AGRICULTURAL CHEMISTRY AND GEOLOGY.
A New Edition, revised and brought down to the present time, by G. T. Atkinson, B.A., F.C.S., Clifton College. Foolscap, 6s. 6d.

Crown 8vo, pp. 760, 7s. 6d.,

AN ETYMOLOGICAL AND PRONOUNCING

DICTIONARY
OF
THE ENGLISH LANGUAGE.

INCLUDING A VERY COPIOUS SELECTION OF

SCIENTIFIC, TECHNICAL, AND OTHER TERMS AND PHRASES.

DESIGNED FOR USE IN SCHOOLS AND COLLEGES,

AND AS

A HANDY BOOK FOR GENERAL REFERENCE.

By the Rev. JAMES STORMONTH,

AND THE

Rev. P. H. PHELP, M.A.

OPINIONS OF THE PRESS.

"This will be found a most admirable and useful Dictionary by the student, the man of business, or the general inquirer. Its design is to supply a full and complete pronouncing, etymological, and explanatory Dictionary of the English language; and, as far as we can judge, in that design it most completely succeeds. It contains an unusual number of scientific names and terms, English phrases, and familiar colloquialisms; this will considerably enhance its value to the general searcher after information. The author seems to us to have planned the Dictionary exceedingly well. The Dictionary words are printed in bold black type, and in single letters, that being the form in which words are usually presented to the reader. Capital letters begin such words only as proper names, and others which are always so printed. They are grouped under a leading word, from which they may be presumed naturally to fall or be formed, or singly follow in alphabetical order—only so, however, when they are derived from the same leading root, and when the alphabetical order may not be materially disturbed. The roots are enclosed within brackets, and for them the works of the best and most recent authorities seem to have been consulted. The meanings are those usually given, but they have been simplified

OPINIONS—*continued.*

as much as possible. Nothing unnecessary is given; but, in the way of definition, there will be found a vast quantity of new matter. The phonetic spelling of the words has been carefully revised by a Cambridge graduate—Mr Phelp; and Dr Page, the well-known geologist, has attended to the correctness of the various scientific terms in the book. The Dictionary altogether is very complete."—*Greenock Advertiser.*

"This Dictionary is admirable. The etymological part especially is good and sound. We have turned to 'calamity,' 'forest,' 'poltroon,' and a number of other crucial words, and find them all derived according to the newest lights. There is nothing about 'calamus,' and 'foris,' and 'pollice truncus,' such as we used to find in the etymological dictionaries of the old type. The work deserves a place in every English School, whether boys' or girls'."—*Westminster Review.*

"A good dictionary to people who do much writing is like a life-belt to people who make ocean voyages: it may, perhaps, never be needed, but it is always safest to have one at hand. This use of a dictionary, though one of the humblest, is one of the most general. For ordinary purposes a very ordinary dictionary will serve; but when one has a dictionary it is well to have a good one. That which is now before us is evidently a work on which enormous pains have been bestowed. The compilation and arrangement give evidence of laborious research and very extensive scholarship. Special care seems to have been bestowed on the pronunciation and etymological derivation, and the 'root-words' which are given are most valuable in helping to a knowledge of primary significations. All through the book are evidences of elaborate and conscientious work, and any one who masters the varied contents of this dictionary will not be far off the attainment of the complete art of 'writing the English language with propriety,' in the matter of orthography at any rate."—*Belfast Northern Whig.*

"When Mr Stormonth tells us that his Dictionary is the result of the labours of many years, we feel certain that he speaks the simple literal truth. Such a work as he has produced required an amount of time, research, toil, and erudition beyond calculation; but we can tell him in return that if to achieve what he undertook compensates him for his pains, he ought to be content, for his volume is really what he designed it to be, a full and complete etymological and explanatory dictionary of the English language. . . . We have not space to describe all its excellences, or to point out in detail how it differs from other lexicons; but we cannot with justice omit mentioning some of its more striking peculiarities. In the first place, it is comprehensive, including not only all the words recognised by the best authorities as sterling old English, but all the new coinages which have passed into general circulation, with a

OPINIONS—*continued.*

great many scientific terms, and those which come under the designation of slang. . . . The pronunciation is carefully and clearly marked in accordance with the most approved modern usage, and in this respect the dictionary is most valuable and thoroughly reliable. As to the etymology of words, it is exhibited in a form that fixes itself upon the memory, the root-words showing the probable origin of the English words, their primary meaning, and their equivalents in other languages. Much useful information and instruction relative to prefixes, postfixes, abbreviations, and phrases from the Latin, French, and other languages, &c., appropriately follow the dictionary, which is throughout beautifully and most correctly printed."—*Civil Service Gazette.*

"This strikes us as likely to prove a useful and valuable work. . . . The number of scientific terms given is far beyond what we have noticed in previous works of this kind, and will in great measure render special dictionaries superfluous. Great care seems also to have been exercised in giving the correct etymology and pronunciation of words. We trust the work may meet with the success it deserves."—*Graphic.*

"This is a work of sterling and rare merit. It must assuredly, and at no distant date, be recognised as the most useful, the most trustworthy, and the most comprehensive of existing cheap dictionaries. We miss in it no feature for which one looks in a dictionary, and we find some which honourably distinguish it from other works of the same kind—for example, phrases are very fully given. The words are printed in large black type, are respelt phonetically, and the etymologies and meanings are given from the latest and best authorities."—*Aberdeen Journal.*

"We feel bound to accord high praise to this work, and we do so with great pleasure. It is extremely suitable for students at college, or in the higher classes of schools; and we know of none of its kind better for general reference, to lie at hand, and be ready for consultation in the study or on the parlour-table."—*Edinburgh Courant.*

" A better guide to the origin, meaning, and pronunciation of the English language has never been published. It has been manifestly the labour of many years to bring it to its present perfection, and it proves, so far as we have yet tested it, so full, and indeed exhaustive, that it is sure to become a standard of reference even amongst expert literati; and at the same time it is so minute, so clear, and so carefully compiled, that it may safely be confided in as expositor and instructor for the beginners in English composition, whether for the press, for private and business correspondence, or for oratorical purposes."—*Newcastle Chronicle.*

NOW PUBLISHING.

ANCIENT CLASSICS
FOR
ENGLISH READERS
BY VARIOUS AUTHORS.

EDITED BY

REV. W. LUCAS COLLINS, M.A.

Author of 'Etoniana,' 'The Public Schools,' &c.

OPINIONS OF THE PRESS.

"We gladly avail ourselves of this opportunity to recommend the other volumes of this useful series, most of which are executed with discrimination and ability."—*Quarterly Review.*

"These Ancient Classics have, without an exception, a twofold value. They are rich in literary interest, and they are rich in social and historical interest. We not only have a faithful presentation of the stamp and quality of the literature which the master-minds of the classical world have bequeathed to the modern world, but we have a series of admirably vivid and graphic pictures of what life at Athens and Rome was. We are not merely taken back over a space of twenty centuries, and placed immediately under the shadow of the Acropolis, or in the very heart of the Forum, but we are at once brought behind the scenes of the old Roman and Athenian existence. As we see how the heroes of this 'new world which is the old' plotted, intrigued, and planned; how private ambition and political partisanship were dominant and active motives then as they are now; how the passions and the prejudices which reign supreme now reigned supreme then; above all, as we discover how completely many of what we may have been accustomed to consider our most essentially modern thoughts and sayings have been anticipated by the poets and orators, the philosophers and historians, who drank their inspiration by the banks of Ilissus or on the plains of Tiber, we are prompted to ask whether the advance of some twenty centuries has worked any great change in humanity, and whether, substituting the coat for the toga, the park for the Campus Martius, the Houses of Parliament for the Forum, Cicero might not have been a public man in London as well as an orator in Rome?"—*Morning Advertiser.*

"A series which has done, and is doing, so much towards spreading among Englishmen intelligent and appreciative views of the chief classical authors."—*Standard.*

"To sum up in a phrase our sincere and hearty commendation of one of the best serial publications we have ever examined, we may just say that to the student and the scholar, and to him who is neither scholar nor student, they are simply priceless as a means of acquiring and extending a familiar acquaintance with the great classic writers of Greece and Rome."—*Belfast Northern Whig.*

List of the Volumes published.

I.—HOMER: THE ILIAD.
By the Editor.

"We can confidently recommend this first volume of 'Ancient Classics for English Readers' to all who have forgotten their Greek and desire to refresh their knowledge of Homer. As to those to whom the series is chiefly addressed, who have never learnt Greek at all, this little book gives them an opportunity which they had not before—an opportunity not only of remedying a want they must often have felt, but of remedying it by no patient and irksome toil, but by a few hours of pleasant reading."—*Times*.

II.—HOMER: THE ODYSSEY.
By the Editor.

"Mr Collins has gone over the 'Odyssey' with loving hands, and he tells its eternally fresh story so admirably, and picks out the best passages so skilfully, that he gives us a charming volume. In the 'Odyssey,' as treated by Mr Collins, we have a story-book that might charm a child or amuse and instruct the wisest man."—*Scotsman*.

III.—HERODOTUS.
By George C. Swayne, M.A.

"This volume altogether confirms the highest anticipations that were formed as to the workmanship and the value of the series."—*Daily Telegraph*.

IV.—THE COMMENTARIES OF CÆSAR.
By Anthony Trollope.

"We can only say that all admirers of Mr Trollope will find his 'Cæsar' almost, if not quite, as attractive as his most popular novel, while they will also find that the exigencies of faithful translation have not been able to subdue the charm of his peculiar style. The original part of his little book—the introduction and conclusion—are admirably written, and the whole work is quite up to the standard of its predecessors, than saying which, we can give no higher praise."—*Vanity Fair*.

V.—VIRGIL.
By the Editor.

"Such a volume cannot fail to enhance the reputation of this promising series, and deserves the perusal of the most devoted Latinists, not less than of the English readers for whom it is designed."—*Contemporary Review*.

"It would be difficult to describe the 'Æneid' better than it is done here, and still more difficult to find three more delightful works than the 'Iliad,' the 'Odyssey,' and the 'Virgil' of Mr Collins."—*Standard*.

VI.—HORACE.

By Theodore Martin.

"Though we have neither quoted it, nor made use of it, we have no hesitation in saying that the reader who is wholly or for the most part unable to appreciate Horace untranslated, may, with the insight he gains from the lively, bright, and, for its size, exhaustive little volume to which we refer, account himself hereafter familiar with the many-sided charms of the Venusian, and able to enjoy allusions to his life and works which would otherwise have been a sealed book to him."—*Quarterly Review.*

"We wish, after closing his book, to be able to read it again for the first time; it is suited to every occasion; a pleasant travelling companion; welcome in the library where Horace himself may be consulted; welcome also in the intervals of business, or when leisure is abundant."—*Edinburgh Review.*

"In our judgment, no volume (of the series) hitherto has come up to the singular excellence of that now under consideration. The secret of this is, that its author so completely puts himself in Horace's place, scans the phases of his life with such an insight into the poet's character and motives, and leaves on the reader's mind so little of an impression that he is following the attempts of a mere modern to realise the feelings and expressions of an ancient. Real genius is a freemasonry, by which the touch of one hand transmits its secret to another; and a capital proof of this is to be found in the skill, tact, and fellow-feeling with which Mr Martin has executed a task, the merit and value of which is quite out of proportion to the size and pretensions of his volume."—*Saturday Review.*

VII.—ÆSCHYLUS.

By Reginald S. Copleston, B.A.

"A really delightful little volume."—*The Examiner.*

"The author with whom Mr Copleston has here to deal exemplifies the advantage of the method which has been used in this series. . . . Mr Copleston has apprehended this main principle, as we take it to be, of his work: has worked it out with skill and care, and has given to the public a volume which fulfils its intention as perfectly as any of the series."—*Spectator.*

VIII.—XENOPHON.

By Sir Alexander Grant, Bart.,
Principal of the University of Edinburgh.

"Sir Alexander Grant tells the story of Xenophon's life with much eloquence and power. It has evidently been with him a labour of love; while his wide reading and accurate scholarship are manifest on nearly every page."—*The Examiner.*

IX.—CICERO.
By THE EDITOR.

"No charm of style, no facility and eloquence of illustration, is wanting to enable us to see the great Roman advocate, statesman, and orator, in the days of Rome's grandeur, in the time of her first fatal hastening to her decadence, with whom fell her liberty two thousand years ago. The first lines of introduction to this fascinating book are full of help and light to the student of the classical times who has not mastered the classical literature, and in whose interests this book is done, simply to perfection."—*Saunders' News-Letter.*

X.—SOPHOCLES.
By CLIFTON W. COLLINS, M.A.

"Sophocles has now been added to the acceptable and singularly equal series of 'Ancient Classics for English Readers.' Mr Collins shows great skill and judgment in analysing and discriminating the plays of the sweet singer of Colonus."—*Guardian.*

XI.—PLINY'S LETTERS.
By the Rev. ALFRED CHURCH, M.A.,
AND
The Rev. W. J. BRODRIBB, M.A.

"This is one of the best volumes of the series called 'Ancient Classics for English Readers.' . . . This graceful little volume will introduce Pliny to many who have hitherto known nothing of the Silver Age."—*Athenæum.*

"Mr Lucas Collins's very useful and popular series has afforded a fit opportunity for a sketch of the life and writings of the younger Pliny; and the writers of the volume before us have contrived, out of their intimate and complete familiarity with their subject, to place the man, his traits of character, his friends, and his surroundings so vividly before us, that a hitherto shadowy acquaintance becomes a distinct and real personage."—*Saturday Review.*

XII.—EURIPIDES.
By W. B. DONNE.

Other Authors, by various contributors, are in preparation.

A Volume is published Quarterly, price 2s. 6d.

W. BLACKWOOD & SONS, 45 GEORGE STREET, EDINBURGH;
AND 37 PATERNOSTER ROW, LONDON.

www.ingramcontent.com/pod-product-compliance
Lightning Source LLC
Chambersburg PA
CBHW020859230426
43666CB00008B/1245